Metallic Biomaterial Interfaces

Edited by
Jürgen Breme, C. James Kirkpatrick,
and Roger Thull

Further Reading

C.S.S.R. Kumar (Ed.)

Nanotechnologies for the Life Sciences

10 Volumes
ISBN 978-3-527-31301-3

N. Kockmann (Ed.)

Micro Process Engineering

Fundamentals, Devices, Fabrication, and Applications

2006
ISBN 978-3-527-31246-7

M. Birkholz

Thin Film Analysis by X-Ray Scattering

2006
ISBN 978-3-527-31052-4

C.S.S.R. Kumar, J. Hormes, C. Leuschner (Eds.)

Nanofabrication Towards Biomedical Applications

Techniques, Tools, Applications, and Impact

2005
ISBN 978-3-527-31115-6

K.A. Jackson

Kinetic Processes

**Crystal Growth, Diffusion, and Phase Transitions
in Materials**

2004
ISBN 978-3-527-30694-7

Metallic Biomaterial Interfaces

Edited by
Jürgen Breme, C. James Kirkpatrick,
and Roger Thull

WILEY-VCH Verlag GmbH & Co. KGaA

The Editors

Prof. Dr. Jürgen Breme
Universität des Saarlandes
LH Metallische Werkstoffe
Postfach 1150
66041 Saarbrücken
Germany

Prof. Dr. C. James Kirkpatrick
Johannes-Gutenberg-Universität
LS für Pathologie, Universitätsklinikum
Langenbeckstrasse 1
55101 Mainz
Germany

Prof. Dr. Roger Thull
Universität Würzburg
LS Funktionswerkstoffe der Medizin
und der Zahnheilkunde
Pleicherwall 2
97070 Würzburg
Germany

Library of Congress Card No.: applied for

British Library Cataloguing-in-Publication Data
A catalogue record for this book is available from the British Library.

Bibliographic information published by the Deutsche Nationalbibliothek
Die Deutsche Nationalbibliothek lists this publication in the Deutsche Nationalbibliografie; detailed bibliographic data are available in the Internet at http://dnb.d-nb.de.

© 2008 WILEY-VCH Verlag GmbH & Co. KGaA, Weinheim

Printed in the Federal Republic of Germany
Printed on acid-free paper

Cover Design   Adam Design, Weinheim
Typesetting   Thomson Digital, Noida, India
Printing   Strauss GmbH, Mörlenbach
Bookbinding   Litges & Dopf GmbH, Heppenheim

ISBN: 978-3-527-31860-5

Contents

Metallic Biomaterial Interfaces. Edited by J. Breme, C. J. Kirkpatrick, and R. Thull
Copyright © 2008 WILEY-VCH Verlag GmbH & Co. KGaA, Weinheim
ISBN: 978-3-527-31860-5

Preface

The Priority Program "Interface between Material and Biosystem" was initiated in 1999 by Prof. Dr.-Ing. Jürgen Breme (Materials Sciences), Saarbrücken, Prof. Dr.-Ing. Roger Thull (Physics), Würzburg, Prof. Dr. med. C. James Kirkpatrick (Pathology), Mainz, and Prof. Dr.-Ing. Hartmut Worch (Materials Sciences), Dresden. The program was financed for a total of six years by the German Research Foundation (DFG).

With a view to overcoming persisting clinical problems, the program was centered on the sustainable improvement of biomaterials for long-term implants. Most of all this requires an improvement of those material characteristics that are determined both by the site of implantation and the functional demands. Specifically, this involves the interface between the material, especially its surface, and the biosystem. The interdisciplinary research approach, with the creation of a functioning network between the various project groups, successfully promoted novel biomaterial modifications, targeted effects on the interface and also permitted subsequent biological reactions to be characterized and finally developed for the field of implantology.

Within the scope of the Priority Program the choice was made to concentrate on titanium and those of its alloys suitable for implant applications, including those of beta or near-beta type. In the case of multiphase alloys there was special interest in the specific features of the individual phases. In projects which were concerned with the specific characteristics of the individual phases it was essential to demonstrate that the results obtained permitted a good correlation between material properties and the reactions of the biological system. In achieving a more general extrapolation of the results it was useful to perform parallel experiments on metals which, though similar to biomaterials on the basis of titanium, also form passive layers in the body, but show different electrical and electronic properties. Among these are austenite steels as well as cobalt-based alloys. The results of physicochemical wear and corrosion experiments also served to improve materials used in implants for load-bearing applications, as did relevant studies on the biological effects of degradation products in cell systems.

The results obtained during this priority program permit correlations between the structure of the interface and the effects on the biological system. Thus, the physical characteristics of the biomaterials with respect to the electronic structure, the

Metallic Biomaterial Interfaces. Edited by J. Breme, C. J. Kirkpatrick, and R. Thull
Copyright © 2008 WILEY-VCH Verlag GmbH & Co. KGaA, Weinheim
ISBN: 978-3-527-31860-5

physicochemical properties of the interface as well as the protein structure following surface interactions could be correlated with the influence on specific cellular functions in human model systems. Among the relevant biological phenomena were pathomechanisms of the induction of inflammation as well as immunological reactions. This was achieved using clinically relevant *in vitro* models with cells of human origin.

With suitable surface modifications or coatings it was possible to promote desirable physicochemical properties of the bulk material and inhibit disadvantageous properties. These modifications included a spectrum of physical, chemical as well as biological methods. In numerous cases it was possible to demonstrate a modification-specific influence of the interface on the interaction with the biosystem. However, coatings with a thickness in excess of 1 μm and aimed at separating the material from the biosystem were not part of this priority program. By contrast, the principal goals of the program focused on studies of improving biocompatibility by modulating the elastic modulus of the bulk material and by mechanical structuring of the surface in relation to the achievable biological efficiency of protein interaction with cells.

All partners involved in the Priority Program ''Interface between Material and Biosystem'' wish to express their sincere gratitude to the DFG for this support. Thanks are also due to all of the scientific reviewers under the leadership of Prof. Dr.-Ing. Dietrich Munz, Karlsruhe, and the program director of the DFG, Dr.-Ing. Frank Fischer, Bonn.

The publication of the results in this book form has been kindly undertaken by Wiley-VCH, Weinheim. The results are presented in three parts: Interface Influence of Materials and Surface Modifications; Physical and Physicochemical Surface Characterization; and Biological Characterization of the Interface and Materials-Related Biosystem Reactions. Finally, thanks are also expressed to Dr.-Ing. Bettina Hoffmann, Bayreuth, Dr. rer. nat. Claus Moseke, Würzburg, as well as PD Dr. habil. Barbara Nebe, Rostock, and Dr. rer. nat. Kirsten Peters, Mainz.

Jürgen Breme
C. James Kirkpatrick
Roger Thull

List of Contributors

Peter Angele
University Hospital Regensburg
Department of Trauma Surgery
Franz-Josef-Strauss-Allee 11
93053 Regensburg
Germany

Frank Aubertin
Saarland University
Department of Metallic Materials
Postfach 151150
66041 Saarbrücken
Germany

Ulrich Beck
University of Rostock
Institute of Electronic Appliances
and Circuits
Richard-Wagner-Strasse 31
18119 Rostock-Warnemünde
Germany

Jürgen Böing
Aachen Resonance Entwicklungs
gesellschaft für magnetresonanz-
kompatible
Medizinprodukte GmbH
Gut Steeg 20
52074 Aachen
Germany

Jürgen Breme
Saarland University
Department of Metallic Materials
Postfach 151150
66041 Saarbrücken
Germany

Egle Conforto
University of La Rochelle
Analytical Center Allee de l'Ocean 5
17071 La Rochelle
France

Rainer Detsch
BioCer Entwicklungs-GmbH
Ludwig-Thoma-Strasse 36c
95447 Bayreuth
Germany

Irina Dieser
University of Bayreuth
Biochemistry
Universitätsstrasse 30
95447 Bayreuth
Germany

Metallic Biomaterial Interfaces. Edited by J. Breme, C. J. Kirkpatrick, and R. Thull
Copyright © 2008 WILEY-VCH Verlag GmbH & Co. KGaA, Weinheim
ISBN: 978-3-527-31860-5

Eva Eisenbarth
South Westfalia University of
Applied Sciences
Institute of Computer and Natural
Sciences
Frauenstuhlweg 31
58644 Iserlohn
Germany

Patrick Elter
University Clinic Würzburg
Department of Functional Materials
in Medicine and Dentistry
Pleicherwall 2
97070 Würzburg
Germany

Andrea Ewald
University Clinic Würzburg
Department of Functional Materials
in Medicine and Dentistry
Pleicherwall 2
97070 Würzburg
Germany

Volker Faust
Hemoteq AG
Adenauerstraße
52146 Würselen
Germany

Ralf-Peter Franke
University of Ulm
Central Institute of Biomedical
Engineering
Albert-Einstein-Allee 47
89081 Ulm
Germany
and
GKSS Research Center
Center for Biomaterial Development
Institute of Polymer Research
Kantstrasse 55
14513 Teltow
Germany

Rosemarie Fuhrmann
University of Ulm
Central Institute of Biomedical
Engineering
Albert-Einstein-Allee 47
89081 Ulm
Germany

Richard H. W. Funk
University of Technology Dresden
Medical Faculty Carl Gustav Carus
Department of Anatomy
Fetscherstrasse 74
01307 Dresden
Germany

Uwe Gbureck
Bayerische Julius-Maximilians-
Universität Würzburg
Department for Functional Materials
in Medicine and Dentistry
Pleicherwall 2
97070 Würzburg
Germany

Jürgen Geis-Gerstorfer
University of Tübingen
Department of Prosthodontics
Section Medical Materials and
Technology
Osianderstrasse 2–8
72076 Tübingen
Germany

Michael Gelinsky
Technical University Dresden
Max Bergmann Center of Biomaterials
Institute of Materials Science
Budapester Strasse 27
01069 Dresden
Germany

André Gorbunoff
University of Applied Sciences (FH)
Dresden
Section of Physics
Friedrich-List-Platz 1
01069 Dresden
Germany

Dirk Handtrack
PLANSEE SE
Innovation Services
6600 Reutte
Austria

Klaus Heckmann
University of Regensburg
Institute of Physical and Theoretical
Chemistry
Universitätsstrasse 31
93053 Regensburg
Germany

Andreas Heeren
University of Tübingen
Institute of Applied Physics
Auf der Morgenstelle 10
72076 Tübingen
Germany

Frank Heidenau
BioCer Entwicklungs-GmbH
Ludwig-Thoma-Strasse 36c
95447 Bayreuth
Germany

Wolfgang Henschel
University of Tübingen
Institute of Applied Physics
Auf der Morgenstelle 10
72076 Tübingen
Germany

Manuela Herklotz
Leibniz Institute of Polymer
Research Dresden
Hohe Strasse 6
01069 Dresden
Germany

Hartwig Höcker
RWTH Aachen
Institute of Technical and
Macromolecular Chemistry
Pauwelsstrasse 8
52056 Aachen
Germany

Bettina Hoffmann
Universität Bayreuth
Friedrich-Baur-Research Institute
for Biomaterials
Ludwig-Thoma-Strasse 36c
95440 Bayreuth
Germany

Günter Kamp
Johannes Gutenberg University
Institute of Zoology
Becherweg 9
55099 Mainz
Germany

Dieter Kern
University of Tübingen
Institute of Applied Physics
Auf der Morgenstelle 10
72076 Tübingen
Germany

Bernd Kieback
Technical University Dresden
Fakultät Machinenwesen
Institute of Materials Science
Helmholtzstrasse 7
01062 Dresden
Germany

C. James Kirkpatrick
Johannes Gutenberg University
Institute of Pathology
REPAIR-Lab
Langenbeckstrasse 1
55101 Mainz
Germany

Doris Klee
RWTH Aachen
Institute of Technical and
Macromolecular Chemistry
Pauwelsstrasse 8
52056 Aachen
Germany

Richard Kujat
University of Regensburg
Center for Medical Biotechnology
Josef-Engert-Strasse 9
93053 Regensburg
Germany

Regina Lange
University of Rostock
Institute of Electronic Appliances
and Circuits
Richard-Wagner-Strasse 31
18119 Rostock-Warnemünde
Germany

Georg Lipps
University of Bayreuth
Department of Biochemistry
Universitätsstrasse 30
95447 Bayreuth
Germany

Frank Lüthen
University of Rostock
Biomedical Research Centre
Cell Biology
Schillingallee 69
18057 Rostock
Germany

René Michalek
University Clinic Würzburg
Department of Functional Materials in
Medicine and Dentistry
Pleicherwall 2
97070 Würzburg
Germany

Thomas K. Monsees
University of Technology Dresden
Medical Faculty Carl Gustav Carus
Department of Anatomy
Fetscherstrasse 74
01307 Dresden
Germany

Claus Moseke
University Clinic Würzburg
Department of Functional
Materials in Medicine and Dentistry
Pleicherwall 2
97070 Würzburg
Germany

Frank A. Müller
University of Erlangen-Nürnberg
Department of Materials Science
Henkestrasse 91
91052 Erlangen
Germany

Lenka Müller
University of Erlangen–Nürnberg
Department of Materials Science
Henkestrasse 91
91052 Erlangen
Germany

Margit Müller
Saarland University
Department of Metallic Materials
Postfach 151150
66041 Saarbrücken
Germany

Rainer Müller
University of Regensburg
Institute of Physical and Theoretical
Chemistry
Universitätsstrasse 31
93053 Regensburg
Germany

J. G. Barbara Nebe
University of Rostock
Department of Internal Medicine
Biomedical Research Centre
Cell Biology
Schillingallee 69
18057 Rostock
Germany

Michael Nerlich
University Hospital Regensburg
Department of Trauma Surgery
Franz-Josef-Strauss-Allee 11
93053 Regensburg
Germany

Kirsten Peters
University of Rostock
Junior Research Group
Department of Cell Biology
Schillingallee 69
18057 Rostock
Germany

Friederike Pfeiffer
University of Bern
Theodor Kocher Institute
Freiestrasse 1
3012 Bern
Switzerland

Tilo Pompe
Leibniz Institute of Polymer Research
Dresden
Hohe Strasse 6
01069 Dresden
Germany

Wolfgang Pompe
Technical University Dresden
Max Bergmann Center of Biomaterials
and Institute of Materials Science
Hallwachsstrasse 3
01062 Dresden
Germany

Gert Richter
Technical University Dresden
Medical Faculty University Hospital
Carl Gustav Carus
Department of Prosthetic Dentistry
01062 Dresden
Germany

Joachim Rychly
University of Rostock
Biomedical Research Centre
Cell Biology
Schillingallee 69
18057 Rostock
Germany

Christa Sauer
Technical University Dresden
Fakultät Maschinenwesen
Institute of Materials Science
Helmholtzstrasse 7
01062 Dresden
Germany

Lutz Scheideler
University of Tübingen
Department of Prosthodontics
Section Medical Materials and
Technology
Osianderstrasse 2–8
72076 Tübingen
Germany

Harald Schmidt
Laboratory for Applied Molecular
Physiology
Becherweg 11
55099 Mainz
Germany

Ulrich T. Seyfert
Institute for blood-research (BluFI)
Eisenbahnstrasse 70
66117 Saaebrücken
Germany

Andrei P. Sommer
University of Ulm
Institute of Micro and Nanomaterials
Albert-Einstein-Allee 47
89081 Ulm
Germany

David Tebbe
Bayerische Julius-Maximilians-
Universität Würzburg
Department for Functional
Materials in Medicine and Dentistry
Pleicherwall 2
97070 Würzburg
Germany

Roger Thull
Bayerische Julius-Maximilians-
Universität Würzburg
Department of Functional
Materials in Medicine and Dentistry
Pleicherwall 2
97070 Würzburg
Germany

Roman Tsaryk
Johannes Gutenberg-University
Institute of Pathology
Langenbeckstrasse 1
55101 Mainz
Germany

Ronald E. Unger
Johannes Gutenberg University
Institute of Pathology
REPAIR-Lab
Langenbeckstrasse 1
55101 Mainz
Germany

Dirk Velten
Saarland University
Department of Metallic Materials
Postfach 151150
66041 Saarbrücken
Germany

Carsten Werner
Leibniz Institute of Polymer Research
Hohe Strasse 6
01069 Dresden
Germany

Stefan Winter
Saarland University
Department of Metallic Materials
Postfach 151150
66041 Saarbrücken
Germany

Hartwig Wolburg
University of Tübingen
Institute of Pathology
Liebermeisterstrasse 8
72076 Tübingen
Germany

Hartmut Worch
Technical University Dresden
Max-Bergmann-Centre of Biomaterials
01062 Dresden
Germany

Günter Ziegler
University of Bayreuth
Friedrich-Baur-Research Institute
for Biomaterials
Ludwig-Thoma-Strasse 36c
95440 Bayreuth
Germany

Andreas Zoll
University Clinic Würzburg
Department of Functional Materials in
Medicine and Dentistry
Pleicherwall 2
97070 Würzburg
Germany

I
Interface Influence of Materials and Surface Modification

1
Introduction

Jürgen Breme

Materials used for the construction of implants have to fulfill mechanical, physical, chemical and biological requirements. Beside suitable mechanical properties like high fatigue strength, sufficient deformatibility and high elasticity they must have biocompatibility and corrosion resistance. The titanium materials meet these requirements best as compared to other metallic biomaterials. Owing to the ever present surface oxide layer with its beneficial thermodynamic, physical and chemical properties, titanium and its alloys are the metallic biomaterials of preference. In particular the strongly negative heat of formation of titanium oxides, which also exists for the primary corrosion products with their large surfaces, provides the corrosion resistance, biocompatibility and low dissolution in body fluids of these materials. In addition to suitable mechanical properties, titanium and its alloys exhibit bioadhesion because of the presence of free hydroxyl groups on the oxide surface [1–3].

Another advantage of the titanium materials is their high elasticity. Young's modulus of these materials is only about 50% of that of CoCr alloys and stainless steel. Because of this fact the so-called stress shielding combined with bone resorption, which occurs in an implant with high stiffness, for the most part can be avoided. Because of a high load transfer new bone formation can be stimulated. Because of this fact the group of β-titanium alloys, which possesses beside a lower notch sensitivity an even lower Young's modulus as compared to cp-Ti or $(\alpha + \beta)$ alloys, attracts more attention and interest. Examples for recently developed β- and near-β-alloys are Ti12Mo6Zr2Fe [4], Ti15Mo5Zr3Al [5], Ti13Nb13Zr [6], Ti30Nb and Ti30Ta [7].

Since the initially existing problems of metallic biomaterials related to mechanical strength properties and corrosion resistance are overcome by the use of titanium materials, actual research can concentrate on the implant/tissue interface in order to improve local biocompatibility. In principle, there exist two possibilities to influence this interface:

1. Structural surface modification in order to change the surface topology (macroscopically, microscopically and nanoscopically), e.g., by means of mechanical, plasma, laser, lithographic or etching procedures.

Metallic Biomaterial Interfaces. Edited by J. Breme, C. J. Kirkpatrick, and R. Thull
Copyright © 2008 WILEY-VCH Verlag GmbH & Co. KGaA, Weinheim
ISBN: 978-3-527-31860-5

2. Chemical surface modification (biological or nonbiological), e.g., by means of thermal and anodic oxidation, sol–gel apatite coatings and covalently or non-covalently attached biomolecules.

Another research effort is occupied with overcoming the disadvantage of the titanium materials of a low wear resistance which is generated by the tendency of cold welding of titanium. Consequently many investigations are engaged in the development of wear-resistant surface layers, e.g., oxides [8], nitrides [9], oxynitrides [10] and diamond-like coatings [11], and in the development of new titanium-based alloys [12].

The aim of the present research project is the improvement of the implant/tissue interface. Besides the development of new abrasion-resistant titanium alloys and coatings, the main goal is an improvement of the interface by means of surface modification.

References

1 Steinemann, S.G. and Perren, S.M. (1985) Titanium as metallic biomaterial, in *Science and Technology* (eds G. Lütjering, U. Zwicker and W. Bunk), DGM Oberursel, pp. 1327–1334.

2 Zitter, H. and Plenk, J. (1987) The electrochemical behavior of metallic implant materials as an indicator of their biocompatibility. *J. Biomed. Mater. Res.*, 21, 881–896.

3 Breme, J. (1989) Titanium and titanium alloys the biomaterials of preference. *Rév. Met.*, 86, 625–637.

4 Trentani, L., Pelillo, F., Pavesi, F.C., Ceciliani, L., Cetta, G. and Forlino, A. (2002) Evaluation of the $TiMo_{12}Zr_6Fe_2$ alloy for orthopaedic implants. *Biomaterials*, 23, 2863–2869.

5 Okazaki, Y. and Ito, Y. (2000) New, Ti alloy without Al and V for medical implants. *Adv. Eng. Mater*, 2 (5), 278–281.

6 Yu, S.Y. and Scully, J.R. Corrosion and passivity of Ti-13% Nb-13% Zr in comparison to other biomedical implant alloys, *Corrosion* 53(12), 965–976.

7 Breme, J. and Wadewitz, V. (1989) Comparison of Ti–Ta, Ti–Nb alloys. *J. Oral. Max. Implants*, 4, 113–118.

8 Hening, F.F. and Repenning, D. (1995) PE-verschleißmindernde Keramik-Metallverbund-Hüftgelenkkugeln. *Unfallchirurg*, 98, 526–529.

9 Dion, I., Roques, X., More, N., Labrousse, L., Caix, J., Lefebvre, F., Rouais, F., Gautreau, J. and Baquey, Ch. (1993) Ex vivo leucocyte adhesion and protein adsorption on TiN. *Biomaterials*, 14(9), 712–719.

10 Thull, R. and Repenning, D. (1990) Funktionelle Beschichtungen für Implantate in Orthopädie und Zahnheilkunde. *Z. Biom. Technik.*, 25, 56–61.

11 Heinrich, G., Rosiwal, S. and Singer, R. F. (1999) in *Symposium 4: Werkstoffe für die Medizintechnik* (eds H. Planck and H. Stallforth), Wiley-VCH, Weinheim, 11–16.

12 Bram, M., Aubertin, F., Venskutonis, A. and Breme, J. (1999) Kinetics of the phase transformation and wear resistance of in-situ processed titanium matrix composites based on Ti-Fe-B. *Mater. Sci. Eng. A*, 264 (1–2), 74–80.

2
Metals and Alloys

2.1
Preparation of Titanium and Titanium Alloys

Stefan Winter, Dirk Velten, and Frank Aubertin

The near-β-alloys Ti13Nb13Zr, Ti30Nb and Ti30Ta and the β-alloy Ti15Mo5Zr3Al were melted to a cigar-like geometry in an electric arc furnace by means of a tungsten electrode in an inert argon atmosphere. To consume the residual oxygen in the furnace atmosphere, a titanium getter was melted before starting the alloy production. The weight ratios of the ingots were adjusted according to the desired alloy composition by the use of the commercially pure metals Ti (grade 2, >99.6 wt.%), Nb (>99.9 wt.%), Zr (>99.9 wt.%), Al (>99.999 wt.%), Mo (>99.95 wt.%) and Ta (>99.8 wt.%). The chemical composition of each alloy was verified by means of energy dispersive X-ray analysis (EDX) (Table 2.1).

After groove rolling of the cigars to a cylindrical shape at 850 °C, the alloys were quenched in water to room temperature. For all alloys, the temperature of 850 °C is high enough to reach the β-phase field. Sample discs with a diameter of 15 mm and a thickness of 2 mm were machined from the groove rolled cylinders. To achieve reproducible surfaces for the oxidation procedure, all samples were ground with SiC paper and mechanically polished with a mixture of a SiO_2 suspension and distilled water. After the metallographic preparation, ultrasonic cleaning in ethanol and subsequent drying, the samples had a mirror-like surface.

Table 2.1 Chemical compositions (wt%) of the alloys determined by means of EDX.

Alloy	Nb	Zr	Mo	Ta	Al	Ti
Ti13Nb13Zr	12.8–13.1	12.7–13.6	0	0	0	bal.
Ti15Mo5Zr3Al	0	4.5–5.2	14.9–15.9	0	2.9–3.1	bal.
Ti30Nb	29.3–29.5	0	0	0	0	bal.
Ti30Ta	0	0	0	31.4–32.3	0	bal.

Metallic Biomaterial Interfaces. Edited by J. Breme, C. J. Kirkpatrick, and R. Thull
Copyright © 2008 WILEY-VCH Verlag GmbH & Co. KGaA, Weinheim
ISBN: 978-3-527-31860-5

Phase Composition

The thermomechanical processing described in Part II resulted (related to the alloy) in the α-and α′-martensitic phase and in the β-phase (near-β-alloys) or only in the β-phase (β-alloys). The presence of these phases was verified by X-ray diffractometry. The phases detected for the four different alloys are listed in Table 2.2 together with the molybdenum equivalency (%Mo_{equiv}) which is defined by Eq. (2.1) and which provides information about the efficacy of different β-stabilizing elements [1]:

$$Mo_{equiv} = [Mo] + 0.2 \times [Ta] + 0.28 \times [Nb] + 0.4 \times [W]$$
$$+ 0.67 \times [V] + 1.25 \times [Cr] + 1.25 \times [Ni] + 1.7 \times [Mn] \qquad (2.1)$$
$$+ 1.7 \times [Co] + 2.5 \times [Fe]$$

where the square brackets represent wt%.

Table 2.2 Phases detected in the alloys by means of X-ray diffraction and the Mo and Al equivalency of the alloys (hcp = hexagonal close-packed; bcc = body-centered cubic).

Alloy	Detected phases	Mo_{equiv}	Al_{equiv}
Ti13Nb13Zr	α′ (martensitic hcp) + β (bcc)	3.64	2.21
Ti15Mo5Zr3Al	β (bcc)	15	4.4
Ti30Nb	α′ (martensitic hcp) + β(bcc)	8.3	–
Ti30Ta	α′ (martensitic hcp) + β (bcc)	6	–

According to other publications, a Mo_{equiv} of 10% is sufficient for a complete stabilization of the β-phase [2]. With decreasing Mo equivalency and with an increasing Al equivalency, which is given in Eq. (2.2) [1], besides the β-phase, the α- and α′-phase also occurs:

$$Al_{equiv} = [Al] + 0.17 \times [Zr] + 0.33 \times [Sn] + 10 \times [O] \qquad (2.2)$$

As a result of the heating to the β-phase field, the microstructures of all alloys are coarse-grained (Figure 2.1). Ti13Nb13Zr (Figure 2.1a) shows the typical appearance of martensitic laths (α′) in a β-matrix as is known after rapid cooling from the β-phase field. Ti15Mo5Zr3Al (Figure 2.1b) consists entirely of equiaxed β-phase grains. With Ti30Nb (Figure 2.1c) a martensitic transformation starting from the grain boundaries can be observed. For Ti30Ta (Figure 2.1d) which has a higher portion of the α′-phase as compared to Ti30Nb (because of the lower coefficient in the Mo equivalency), the martensitic α′-phase which is embedded in a β-matrix is much finer and more acicular than the martensite in Ti13Nb13Zr and in TiNb30 because of a lower martensite starting temperature.

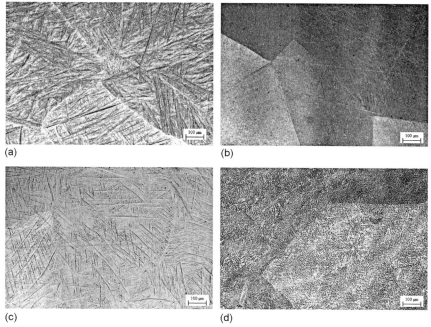

Figure 2.1 Microstructure of the β- and near-β-titanium alloys:
(a) Ti13Nb13Zr; (b) Ti15Mo5Zr3Al, (c) Ti30Nb; (d) Ti30Ta.

2.2
Development of Wear-Resistant Nanocrystalline and Dispersion Strengthened Titanium Materials for Implants

Christa Sauer, Dirk Handtrack, and Bernd Kieback

The disadvantages of cp-Ti as an implant material are its low strength and low wear resistance. Until recently, the improvements of strength and wear resistance of titanium have been realized by alloying and various wear-resistant coatings. Alloying has severe drawbacks due to the potential cell toxicity effects of many alloying elements (e.g. V, Co, Cr). Coatings have flake-off issues which make them unsuitable in real applications.

The aim of the present investigations was to increase the strength and the hardness by using the mechanisms of grain size refinement and dispersion strengthening (ds). Titanium silicides and titanium carbides are well suited as dispersoids, because they do not influence the protective passive layer of titanium. Furthermore, silicon and carbon are not harmful to human body tissues.

The manufacture of the desired microstructure was carried out with powder metallurgy processes. For this purpose, mixtures of titanium powder with the addition of silicon powder, graphite powders or metalorganic fluid hexamethyldisilane (HMDS) were high-energy ball milled. Using this method, nanocrystalline powder granules with fine distributions of the dispersoid-forming elements silicon and/or carbon were produced. To prevent grain coarsening during sintering, the novel technique of spark plasma sintering (SPS) was applied [3]. The microstructure of the compacted materials comprises an ultrafine-grained (ufg) titanium matrix with an average grain size of 200–500 nm and embedded dispersoids in the range of 50–300 nm (Figure 2.2). The dispersoids were identified as titanium silicides Ti_5Si_3 and/or titanium carbides TiC_x as result of solid-state reaction between the dispersoid-forming agents Si, C or HMDS with titanium.

As shown in Table 2.3, it was possible to increase the material strength and hardness in comparison to cp-Ti by producing an ultrafine titanium matrix grain size along with the presence of dispersoids. The mechanical properties were even better than that of TiAl6V4 alloy. A sufficient plastic elongation was only realized by the composition Ti/1.3HMDS (wt%). The excellent deformation behavior of this material is the result of a bimodal grain size distribution caused by a low dispersoid content and an inhomogeneous dispersoid distribution [4]. Figure 2.3 shows that the wear resistance of titanium was improved clearly by the increase of the hardness and

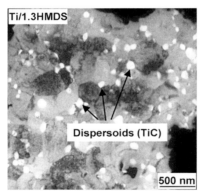

Figure 2.2 Transmission electron micrograph showing the microstructure of Ti/1.3HMDS and the TiC dispersoids distribution.

Table 2.3 Mechanical properties of the newly developed materials in comparison to cp-Ti and TiAl6V4.

	Reference materials		ufg and ds Ti			
	cp-Ti	TiAl6V4	Ti/1Si	Ti/1.5C	Ti/1.3HMDS	Ti/2.6HMDS
Hardness (HV0.5)	160	300	360	350	320	350
Yield strength (MPa)	480	1220	1420	1135	1310	1660
Bending strength (MPa)	860	2090	1650	1620	1930	1980
Plastic elong. at break (%)	6.1	9.6[a]	0.5	2.0	0.8	9.6

[a] Termination of bending test.

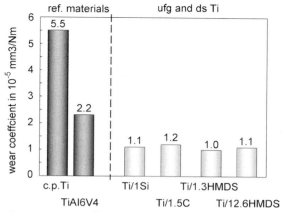

Figure 2.3 Wear coefficients of the newly developed materials in comparison to cp-Ti and TiAl6V4. Results are for pin-on-disc wear tests; the disc material was TiAl6V4.

strength of the newly developed material. The wear resistance of the new developed materials was twice as good as that of TiAl6V4.

To characterize the biocompatibility of the wear residue originating from the tribological tests, its interaction with human endothelial cells was investigated, in particular with regard to pro-inflammatory reactions and cell vitality. To prevent influence of the alloying elements from the disc material (e.g. vanadium from TiAl6V4 discs), wear particles produced from cp-Ti discs were used. As shown in Figure 2.4, the interaction of the wear with the endothelial cells is minimal or

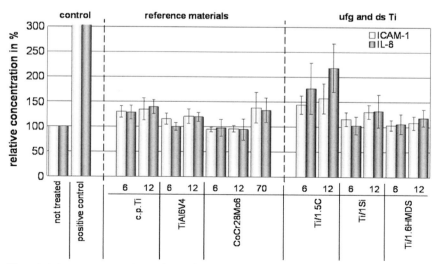

Figure 2.4 Concentration and IL-8 release after 24 h of exposure in different particle concentrations in μg ml^{-1} (positive control IL-8 ≈ 1600%, ICAM-1 ≈ 1200%).

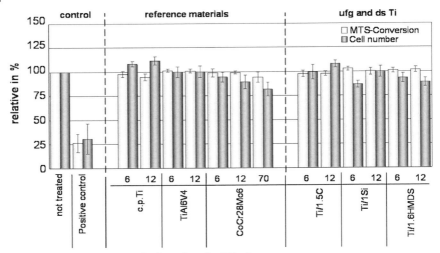

Figure 2.5 MTS conversion and cell number after 72 h of exposure in different particle concentrations in $\mu g\, ml^{-1}$.

negligible, and concentration dependence increased in response to the intake of particles. The effects are insignificantly higher for the wear particles of Ti/1.5C (wt%) than for all other particles tested (up to 2.2-fold increase). The vitality of the cells after 72 h exposure is not significantly affected; only the concentration-dependent reduction of cell number induced by CoCr28Mo6 wear appears to be relevant (Figure 2.5).

To estimate the suitability of the newly developed materials for implant application, it was also necessary to investigate their corrosion behavior (in particular with regard to pitting in the presence of chloride ions) and the semiconductivity (n-type) of their passive layers in comparison to cp-Ti. Potentiodynamic measurements in phosphate buffered saline solution (PBS) at 37 °C showed that the breakthrough potentials U_B are clearly higher than the normal redox potential of biological systems ($U_{Redox} = 350$–$550\, mV$ referred to the saturated calomel electrode (SCE) [5]). The corrosion resistance of the new materials is much higher than that of TiAl6V4 and slightly lower than that of cp-Ti (Figure 2.6). The small decrease of U_B in comparison to cp-Ti is probably the result of the ultrafine-grained microstructure and the dispersoids. The n-type semiconductor properties were characterized with electrochemical impedance spectroscopy (EIS) measurements by calculating the flatband potentials U_{FB} and the donor densities N_D using the Mott–Schottky equation [6]. The values are presented in Table 2.4. The U_{FB} values of the new materials are clearly lower than the normal redox potential of biological systems, and even lower that of TiAl6V4. The presence of silicon or rather Ti_5Si_3 leads to a decrease of the flatband potential below the value of silicon-free titanium. The donor densities are within an order of magnitude of each other: their differences are insignificant. Only TiAl6V4 shows a slightly increased N_D.

Summarizing, the ultrafine-grained and dispersion-strengthened titanium materials can be manufactured by the powder metallurgy processes of high-energy ball

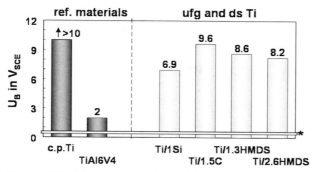

Figure 2.6 Breakthrough potentials U_B of the newly developed materials in comparison to cp-Ti and TiAl6V4; the asterisk indicates the normal redox potential of biological systems (U_{Redox}: 350–550 mV$_{SCE}$) [3].

Table 2.4 Flatband potential and donor density of the new developed materials in comparison to cp-Ti and TiAl6V4.

	Reference materials		ufg and ds Ti			
	cp-Ti	TiAl6V4	Ti/1Si	Ti/1.5C	Ti/1.3HMDS	Ti/2.6HMDS
Flatband potential (mV$_{SCE}$)	110	226	63	196	93	115
Donor density (10^{20} cm^{-3})	3.09	4.49	2.03	2.60	1.72	1.75

milling and compaction via spark plasma sintering. Ti/1.3HMDS (Ti/0.5Si/0.65C) has the best combination of properties for implant applications.

2.3
Amorphous Titanium–Silicon Alloys: New Materials for Implants

Gert Richter, Hartmut Worch, and Wolfgang Pompe

Cp-titanium has proven itself in the form of different implants. However, the abrasion properties which are necessary for hip implants are unsatisfactory. These properties can only be achieved by alloying. A new, previously unused titanium–silicon binary system has been introduced. Two aims were pursued with the introduction of silicon:

1. Keeping the very good biocompatibility of titanium.
2. Improvement of the mechanical properties, particularly resistance to abrasion.

In accordance with the Ti–Si constitution diagram a series of intermetallic phases occur depending on the composition. It can be seen from the titanium–silicon phase diagram that silicon can only be solved by contents of ~0.1 at.% at a temperature of

600 °C. The intermetallic phase Ti_3Si occurs at higher silicon contents. It is to be expected that the phases improve the mechanical properties but not the degradation behavior (corrosion behavior). The structure must be single-phase for a small degradation. The single-phase structure can be achieved as the silicon has the property of glass forming.

This is possible, amongst other methods, by rapid solidification which results in amorphous alloys. These are known for their high tensile strength, high fracture toughness and homogeneous flow at low strain [7]. Surfaces with this kind of high strengths and ductility have better wear properties and tend less to form surface cracks. Amorphous alloys are clearly superior to those in a crystalline state, particularly with respect to local corrosion, e.g. pitting, due to having no grain boundaries and their single-phase state.

Incompatibility symptoms [8] in body tissue caused by silicon have so far not been documented. Surface coatings of amorphous Ti–Si alloys, which were intended for different purposes, have already been produced in the past by magnetron sputtering with silicon proportions of 10–100% [9]. Investigations into the properties of these alloys, particularly with regards to application in implantology, are unknown.

Alloy Development

Amorphous alloys can today mainly be produced in the form of thin films. Laser ablation (cross-beam pulsed laser deposition) and magnetron sputtering are used. With the aid of these processes Ti–Si alloys with compositions of around Ti 70 Si 30, Ti 50 Si 50 and Ti 30 Si 70 were applied in thin films onto titanium. The film thicknesses were 1 µm. The characterization of structure and element distribution was carried out by scanning electron microscopy (SEM), energy dispersive X-ray analysis (EDX), transmission electron microscopy (TEM) and X-ray diffraction (XRD).

All of the films show an even distribution of elements and an amorphous structure. The electron diffraction patterns in Figs. 2.7 and 2.8 demonstrate the results.

Figure 2.7 Bright-field image of amorphous Ti–Si film laser ablation (cross-beam pulsed laser deposition), 38.3 at.% Ti/ 61.7 at.% Si.

Figure 2.8 Dark-field image of amorphous Ti–Si film magnetron sputtering, 69.6 at.% Ti/30.4 at.% Si.

Structure Characterization

The Ti–Si films produced were investigated with regards to their structure with a TEM LEO EM 912 Omega instrument. The structural investigations were carried out by electron beam diffraction and the chemical composition was determined by EDX.

In all cases the diffraction patterns presented an amorphous structure (Figs. 2.7 and 2.8). The bright- and dark-field images clearly show that homogeneous, closed and pore-free films can be obtained from both film production processes.

Investigations on Oxide Film Formation Using X-ray Photoelectron Spectroscopy (XPS) Analyses

It is known that the degradation behavior of alloys depends on the composition and structure of the passive films which are formed on the surface of the alloys. The passive films formed by air contact on the surface of the alloys were analyzed by XPS (PHI 560 CI, FEI). The surface removal rate was $1 \, nm \, min^{-1}$.

The passive films are oxidic and *ca* 3 nm thick. The cation proportion in the oxide of the alloy composition corresponds to the original material composition (Figure 2.9). Crystalline phases and unmixing phenomena were not observed. Depending on alloy composition these contain titanium, silicon, nitrogen and oxygen ions. The film material demonstrates a relatively high level of O and N, each at 15 at.%. These elements are integrated by the high reactivity of the titanium during the deposition of the films out of the residual gas. Titanium cannot be deposited oxygen-free even under ultrahigh vacuum (10^{-10} mbar basis pressure). Even under the conditions mentioned above, 2 to 5 at.% remains. This ensures that nitrogen does not get into the alloy volume during the oxidation process but continually during the deposition process. The carbon detected in the areas near the surface points to possible contamination.

Investigations on the Thermal Stability of Amorphous Ti–Si Films

As friction and abrasion always produce warmth the thermal behavior of the films produced is of interest. From a thermodynamic point of view amorphous alloys are far from thermodynamic equilibrium and thus are metastable. However,

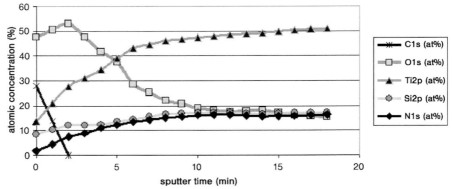

Figure 2.9 XPS analysis: depth profile measurement of an alloy with 70 at.% Ti and 30 at.% Si.

crystallization phenomena were first observed in TEM over *ca* 1000 K during heating tests on amorphous Ti–Si films (Figure 2.10). At temperatures of 123 °C, which are used during steam sterilization, the Ti–Si films remained intact.

Corrosion Resistance (Degradation Behavior)
Potentiodynamic investigations in body-analogous media were carried out in order to demonstrate the corrosion behavior of the amorphous Ti–Si alloys produced. In Figure 2.11 current density–potential curves of cp-Ti and a Ti–Si alloy are shown. DIN saliva (0.1 M NaCl/0.1 M $C_3H_6O_3$, 1 : 1) was used as the corrosive. As can be seen from the curves the amorphous Ti–Si alloy demonstrates excellent resistance to corrosion at 37 °C. Marginally better corrosion behavior can be deduced from the comparatively higher leakage potential of the Ti–Si alloy. The reason for this was found in the XPS analyses which show that silicon ions are integrated into the passive film.

Mechanical-Technological Properties
The characterization of the mechanical-technological properties of amorphous Ti–Si films was carried out by means of universal hardness measurements using a nanoindenter. As Table 2.5 shows, a change in the alloy composition results in a significant change in the hardness. In alloys with 70 at.% Si there is a threefold increase in the hardness when compared to cp-Ti.

Abrasion Test
The pin-on-disc configuration was chosen for abrasion tests. The disc was in the form of cold rolled cp-Ti with a sputtered surface film and the pin was made of titanium. The stresses were chosen so that in accordance with a model for calculating joint forces in a human, surface pressure of 80 N cm^{-2} was produced between the disc and the pin. The tests were carried out in air and in Ringer's solution at 37 °C in order to simulate the biological environment in the body. In Figure 2.12 the surface topographies are shown according to selected load cycles. After 500 cycles the applied film of 1 μm had already been completely abraded. This result was also confirmed by EDX of the abrasion trace.

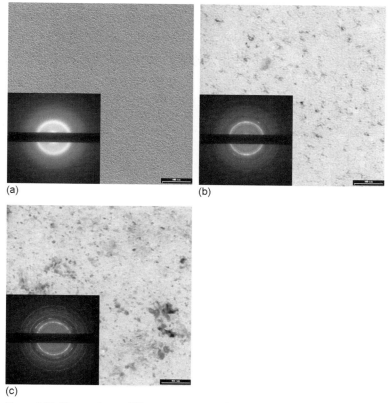

(a)

(b)

(c)

Figure 2.10 Electron beam diffraction patterns and bright field images of amorphous Ti–Si films after heating-up tests in TEM (magnetron sputtered sample): (a) original state (RT); (b) sample 1 (1000 K); (c) sample 1 (1165 K).

Higher silicon proportions in the alloy increase the hardness as expected (Table 2.5) and should impart higher resistance to abrasion as a result. However, for thin films such behavior can only then be expected when the film is strongly bonded onto the bulk material. This was not the case according to the experimental findings.

There is therefore no doubt that the previously chosen production process with regards to a practical application of amorphous Ti–Si alloys is not suitable. However, the current state of knowledge on mechanical behavior opens up the possibility of removing the cause of the inadequate adhesive strength, for example with gradient layers.

Physiological Cell Investigations
Primary osteoblasts from rat calvaria were cultivated on films in cell adhesion investigations. In all the samples the number of cells increases with growing cultivation duration. However, only a small proportion of cells are actin positive to start with. This proportion increases within the first 4 h of cultivation and remains constant thereafter until the end of the test (24 h). The proportion is approximately

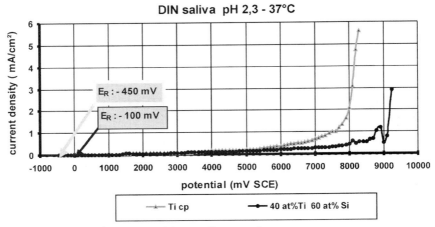

Figure 2.11 Current density–potential curve of an amorphous Ti–Si alloy (40 at.% Ti/60 at.% Si) and cp-Ti.

Table 2.5 Universal hardness of amorphous Ti–Si films in comparison to pure titanium.

Alloy	A	B	C	D	E	cp-Ti
Composition	70 at.% Ti 30 at.% Si	60 at.% Ti 40 at.% Si	50 at.% Ti 50 at.% Si	40 at.% Ti 60 at.% Si	30 at.% Ti 70 at.% Si	99.7% Ti
DHV 0.0002	545	549	676	768	774	243

the same in all the Ti–Si alloys. In cp-Ti, however, the proportion of actin-positive cells after 24 h cultivation is marginally lower. This indicates better acceptance of the amorphous surfaces containing silicon in comparison to cp-Ti.

The cause of the improved cell behavior in rat osteoblasts on Ti–Si surfaces has still not been clarified. It is assumed that the silicon ions integrated into the passive film have a positive effect on cell proliferation and on the formation of the actin cytoskeletons. This matter must be clarified in further studies with the aid of proliferation investigations.

As can be seen from Figure 2.13, the enzyme activity of alkali phosphatase is improved through the proportion of silicon in the passive film in comparison to cp-Ti. Ascertaining the cause of this increase requires further investigations.

Within the framework of the program, an alloy system was suggested for the first time that was aiming not only for the improvement of the biocompatibility of titanium but also for increased resistance to abrasion. A binary system was chosen consisting of the components titanium and silicon. Amorphous Ti–Si alloys were produced with varying proportions of silicon. The biocompatibility of these alloys is better than that of cp-Ti. Unfortunately the improvement of resistance to abrasion could not be achieved.

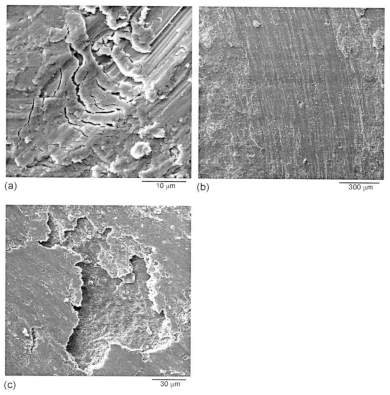

(a) 10 μm   (b) 300 μm

(c) 30 μm

Figure 2.12 Abrasion marks on alloy A (30 at.% Si) after stressing the sample in air depending on the test duration: (a) after 100 cycles; (b) after 500 cycles; (c) after 1000 cycles.

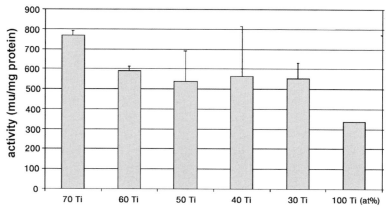

Figure 2.13 Enzyme activity of alkali phosphatase in rat osteoblasts on amorphous Ti–Si alloys and on pure titanium.

References

1 Collings, E.W. (1994) *Titanium Alloys: Materials Properties Handbook* (eds) R. Boyer, G. Welsch and E.W. Collings), ASM International,Materials Park, OH, p. 3.

2 Thull, R. and Repenning, D. (1990) Funktionelle Beschichtungen für Implantate in Orthopädie und Zahnheilkunde. *Z. Biom. Technik.*, **25**, 56–61.

3 Handtrack, D., Despang, F., Sauer, C., Kieback, B., Reinfried, N. and Grin, Y. (2006) Fabrication of ultra-fine grained and dispersion-strengthened titanium materials by spark plasma sintering. *Mater. Sci. Eng. A*, **437** (2), 423–429.

4 Handtrack, D., Sauer, C. and Kieback, B. Microstructure and properties of ultra-fine grained and dispersion-strengthened titanium materials for implants, *J. Mater. Sci.* (accepted).

5 Velten, D., Biehl, V., Aubertin, F., Valeske, B., Possart, W. and Breme, J. (2002) Preparation of TiO_2 layers on cp-Ti and Ti6Al4V by thermal and anodic oxidation and by sol-gel coating techniques and their characterization. *J. Biomed. Mater. Res.*, **59**, 18–28.

6 Scharnweber, D., Beutner, R., Rößler, S. and Worch, H. (2002) Electrochemical behaviour of titanium-based materials: are there relations to biocompatibility? *J. Mater. Sci. Med.*, **13**, 1215–1220.

7 Cahn, R.W. and Greer, A.L. (1996) Metastable states of alloys, in *Physical Metallurgy*, 4th revised and enhanced edn (eds R.W. Cahn and P. Haasen), Elsevier Science, p. 1723.

8 Wirz, J. and Bischoff, H. (1997) *Titan in der Zahnmedizin*, Quintessenz Verlags-GmbH, Berlin, p. 392.

9 Naka, M., Matsui, T., Maeda, M. and Mori, H. (1995) *Mater. Trans., JIM*, **36** (7), 797–801.

3
Topological Surface Modifications

3.1
Patterning by Mechanical Treatment

Bettina Hoffmann, Günter Ziegler, Regina Lange, and Ulrich Beck

The interaction between implant material and biosystems (e.g. protein adsorption, cell adhesion) is markedly influenced by the topography of the implant surface and depends on the specific cell types involved. In many cases, a direct correlation between cell growth *in vitro* and surface roughness has been reported [1–8]. Systematic investigations have been made to evaluate the correlation between surfaces patterning on a micro- and nanometer scale and cell behavior. The surface roughness was adjusted by different techniques and the cell response to the modified surfaces was examined by *in vitro* cell tests (see Part III).

The surface structure of cp-Ti (grade 2) was modified using a broad range of mechanical processes (polishing, machining, blasting with glass beads (at 3.5 bar) and corundum particles (Al_2O_3; at 2.5 and 6 bar)). The various processes resulted in a roughness palette ranging from $R_a = 0.07 \, \mu m$ (polished, Figure 3.1) to $R_a = 6 \, \mu m$ (corundum blasting at 6 bar, Figure 3.5). The machined surfaces (unlike the surfaces obtained with all other methods) had a defined and regular surface structure in the form of concentric circular grooves with a mean roughness ranging from 0.5 to 1.7 µm (Figure 3.2). Glass bead blasting (Figure 3.3) produced a consolidated and leveled material surface ($R_a = 1.2 \, \mu m$). Corundum blasting produced a substantially "fractured" surface with extremely sharp edged structures. The mean roughness values for these specimens ranged from 4 µm (2.5 bar, Figure 3.4) to 6 µm (6 bar, Figure 3.5). For further characterization of the surface properties, see Part II.

In additional to the cp-Ti, we prepared surfaces of different roughness from Ti6Al4V alloy. In order to produce different patterns on the metallic surfaces, we prepared metallic discs ($d = 15 \, mm$) of Ti6Al4V alloy and ground them with SiC sandpaper of decreasing grain sizes (grit sizes P320, P500 and P1000). Other samples were polished with OP-S suspension after grinding with SiC-P1000 sandpaper. The polished and ground surfaces were cleaned using ultrasound, with the following sequence of solvents: tenside, distilled water, cyclohexane, acetone and ethanol.

Metallic Biomaterial Interfaces. Edited by J. Breme, C. J. Kirkpatrick, and R. Thull
Copyright © 2008 WILEY-VCH Verlag GmbH & Co. KGaA, Weinheim
ISBN: 978-3-527-31860-5

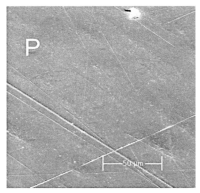

Figure 3.1 SEM image of a polished (P) titanium surface, $R_a = 0.07\,\mu m$.

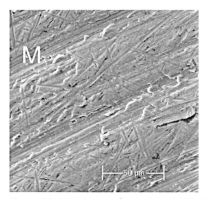

Figure 3.2 SEM image of a machined (M) titanium surface, $R_a = 1.5\,\mu m$.

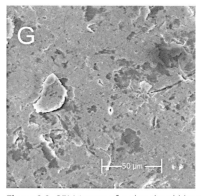

Figure 3.3 SEM image of a glass bead-blasted titanium surface, $R_a = 1.2\,\mu m$.

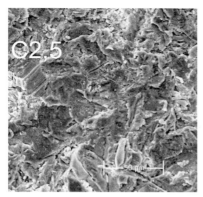

Figure 3.4 SEM image of a titanium surface blasted with corundum at 2.5 bar, $R_a = 4\,\mu m$.

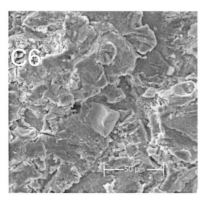

Figure 3.5 SEM image of a titanium surface blasted with corundum at 6 bar, $R_a = 6\,\mu m$.

Grinding produced grooves of different depth and direction. Subsequent TiO_2 coating (see Section 4.1.1) reproduced these grooves (Figures 3.6 and 3.7). The following roughness values were produced by those treatments: $R_a = 0.014\,\mu m$ (polished), $R_a = 0.059\,\mu m$ (P1000 SiC paper), $R_a = 0.090\,\mu m$ (P500 SiC paper) and $R_a = 0.12\,\mu m$ (P320 SiC paper).

In clinical use, load-bearing, cementless implants have a mostly sandblasted surface treated with SiO_2 or Al_2O_3 particles to provide a good mechanical interlock at the interface between implant and bone. In this investigation, we used the sandblasting technique (Al_2O_3 particles; $d = 110\,\mu m$, 4 bar, 25 cm distance) to produce surfaces with a higher roughness ($R_a = 1.01\,\mu m$) than the ground samples. Utilizing this method, particles used for blasting are often found attached to the surface, which have been discussed as a potential cause of abrasion [9]. Figure 3.8a shows a scanning electron microscopy image of a sandblasted surface, which has been cleaned subsequently by ultrasound. Raman spectroscopic measurements show (Figure 3.8b) that Al_2O_3 particles remain attached to the surface.

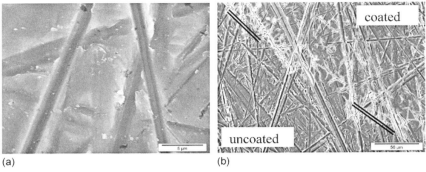

(a)   (b)

Figure 3.6 SEM images of Ti6Al4V discs: (a) ground with P320 SiC sandpaper; (b) TiO_2 coating at the transition between coated and uncoated area of Ti6Al4V disc ground with 320 SiC paper.

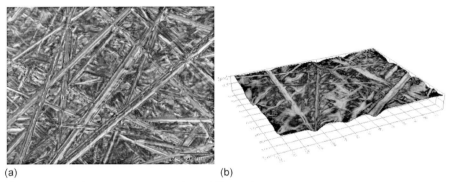

(a)   (b)

Figure 3.7 (a) Light microscopic image (DIC, differential inter-ference contrast) of TiO_2-coated Ti6Al4V ground with P320 SiC paper and (b) three-dimensional calculated surface topography.

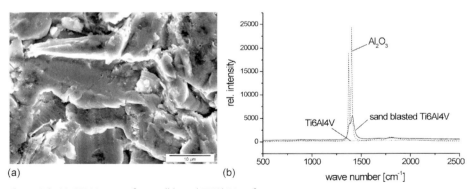

(a)   (b)

Figure 3.8 (a) SEM image of a sandblasted Ti6Al4V surface. (b) Raman spectroscopic measurements of Ti6Al4V, sandblasted Ti6Al4V and Al_2O_3 particles used for sand blasting as reference.

3.2
Etching Techniques

3.2.1
Bioactive Nanostructures

Lenka Müller, Egle Conforto, and Frank A. Müller

The surface state of an implant alters the bone response and fixation [10,11]. Characteristics such as surface composition and topography (i.e. roughness) are of particular importance. These surface features are interrelated and thus it is difficult to differentiate which affects the implant integration into the tissue at most [12,13]. However, the surface roughness was described to influence the amount of adsorbed proteins since the material is immediately conditioned by proteins or other components from the blood and thus the cells do not encounter the original surface *in vivo* [14,15]. There is much evidence that surface roughness enhances osseointegration of titanium implants. Furthermore, the surface roughness influences surface wettability which in turn determines further cellular reaction as well [12,16].

In the present section, the influence of acid etching and subsequent alkali treatment of commercially pure titanium on surface composition, topography and roughness is discussed using transmission electron microscopy and selected area electron diffraction pattern analysis (TEM/SAED), scanning electron microscopy (SEM), X-ray photoelectron spectroscopy (XPS), X-ray diffraction (XRD) and laser scanning microscopy (LSM).

Acid-Etched Microroughened Surfaces

Acid etching of titanium in HCl under inert atmosphere was examined as an alternative pretreatment to obtain a uniform initial titanium surface before alkali treatment. Cp-Ti squares with an edge length of 10 mm and a thickness of 1 mm were washed in isopropanol and distilled water in an ultrasonic cleaner and subsequently dried at 100 °C. After the cleaning procedure the samples were etched in 37% HCl under inert atmosphere of argon at temperatures up to 50 °C for different times.

Figure 3.9 shows SEM micrographs of titanium surfaces after acid etching in HCl for various treatment times and at different temperatures. The surface roughness increases with etching time and reaches a maximum of 4 µm during combined treatment at 50 °C for 90 min and 40 °C for 60 min. The surface roughness shown in Figure 3.10 was measured by LSM. At prolonged etching time and elevated temperature the sample dimensions changed markedly which can in turn have a negative influence on the mechanical properties especially in the case of small or thin-walled implants.

Contact angle measurements using sessile drop experiments reveals the hydrophobic behavior of the etched surface with contact angles of 115°, whereas measurements on cp-Ti showed contact angles of 74°.

XRD was used to identify crystallographic phases (Figure 3.11a). The acid-etched substrate consists of hexagonal α-titanium and titanium hydride (TiH_2). The formation of titanium hydride was confirmed using TEM/SAED on cross-sectional specimens (Figure 3.11b). For this purpose two parts of the sample were cut perpendicular

(a)

(b)

(c)

(d)

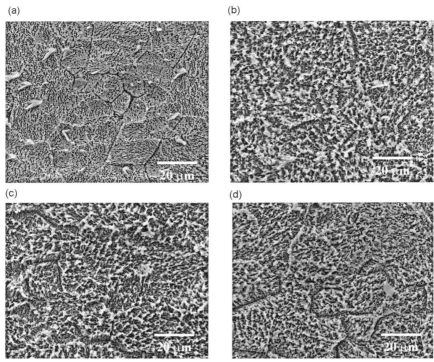

Figure 3.9 SEM micrograph of titanium surface after acid etching at (a) 50 °C/30 min, (b) 50 °C/60 min, (c) 50 °C/90 min and (d) 50 °C/90 min +40 °C/60 min.

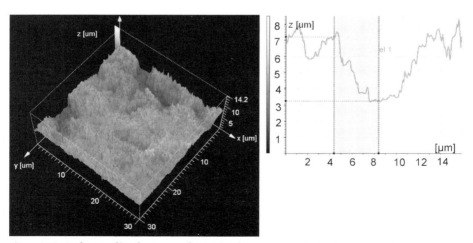

Figure 3.10 Surface profile of titanium after HCl etching at 50 °C/90 min +40 °C/60 min.

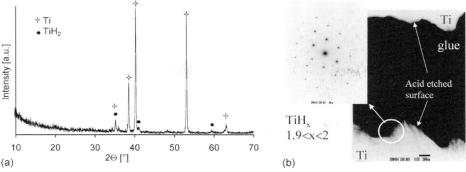

TiH$_x$
1.9<x<2

(a)

(b)

Figure 3.11 (a) XRD and (b) TEM/SAED analysis of titanium after HCl etching at 50 °C/90 min + 40 °C/60 min.

to the surface and sandwiched together with an epoxy glue, surface towards surface. The low-magnification cross-sectional TEM image shows both sides of the specimen separated by the glue (Figure 3.11b). The reflections in the corresponding SAED pattern fit with those of fcc TiH$_{1.971}$ phase (JCPDS card 07-0370) belonging to the space group *Fm3m* with a lattice parameter $a = 4.44$ Å. The data can be associated with several TiH$_x$ phases ($1.9 < x < 2$) that have fcc structure above room temperature [17].

Using XPS, titanium, oxygen, chlorine and carbon were detected in the sample surface (Figure 3.12). Chlorine (up to 10 at.%) resulting from the etching in HCl was only found in the outermost surface up to sputtering depth of 10 nm. The depth of oxygen incorporation and the course of the oxygen sputter profile suggest a diffusion character of oxygen incorporation into the film (Figure 3.12). It was described in the literature that exposure of the TiH$_x$ film to air atmosphere might cause both surface and bulk oxygen interaction [18]. Oxygen interacts with TiH$_x$, replacing hydrogen on the surface and leading to the formation of a thin titanium oxide layer that protects the deeper parts of TiH$_x$ against interaction with O$_2$. A thin titanium oxide layer is

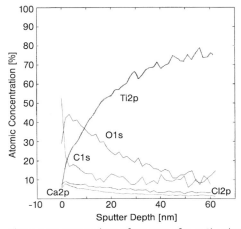

Figure 3.12 XPS analysis of titanium after HCl etching at 50 °C/90 min +40 °C/60 min.

contaminated by carbon monoxide, hydrocarbons and nitrogen. The thickness of this layer ($\sim$15 nm) is lower than the corresponding layer on Ti ($\sim$30 nm). A passivation layer, preventing deeper oxidation of titanium, is probably formed in the TiO_2/TiH_y interface region.

Based on the results described above, possible structural changes of the titanium surface after acid etching are as follows. Titanium is passivated by an oxide film that forms spontaneously. Its composition is reported to consist of an outer TiO_2 layer and an intermediate TiO_x layer [19]. During acid etching in HCl the passive oxide film degrades according to

$$TiO_2 + 4HCl \rightarrow TiCl_4 + 2H_2O \tag{3.1}$$

Simultaneously titanium reacts with HCl to form $TiCl_3$ and H_2:

$$2Ti + 6HCl \rightarrow 2TiCl_3 + 3H_2 \tag{3.2}$$

TiH_2 will be formed according to

$$Ti + H_2 \rightarrow TiH_2 \tag{3.3}$$

acting as an intermediate layer for the formation of a new oxide layer in contact with air.

Alkali-treated Nanostructured Surfaces

Considerable efforts have been made to improve the bone integration of titanium by a modification of the oxide film [20]. Bioactivity and thus bone-bonding ability of a material is generally associated with apatite formation during soaking tests in simulated body fluid (SBF). It was shown that the precipitation of apatite on a titania gel is induced by the abundant $Ti-OH^-$ groups on its surface, implying that a large number of surface $Ti-OH^-$ groups are essential for apatite nucleation [21].

In order to increase the number of $Ti-OH^-$ groups, acid-etched titanium samples (50 °C/90 min +40 °C/60 min) were alkali treated in 10 mmol L^{-1} NaOH at 60 °C for 24 h as described in [22,23].

Figure 3.13 shows SEM micrographs of the titanium surface after acid etching in HCl at 50 °C/90 min + 40 °C/60 min and subsequent alkali treatment in NaOH. EDX analysis revealed titanium, oxygen and sodium to be present in the surface (Figure 3.13a, inset). The surface topography obtained by acid etching is maintained and an additional nano-substructure developed as shown by LSM profile measurements in Figure 3.14. The alkali modification shifts the hydrophobic state of the acid-etched surface to a hydrophilic one showing contact angles of 10°. It can be concluded than the NaOH treatment significantly improves the wettability of the surface by retaining the surface microstructure.

Using XPS oxygen, titanium and about 10 at.% of sodium were detected in the sample surface (Figure 3.15). XRD analysis (spectra not shown) did not reveal any new phases compared to acid-etched samples. Since only reflection of a single hexagonal α-titanium and titanium hydride (TiH_2) were detected, the formation of an amorphous or very thin layer can be assumed.

The titanium oxide layer covering acid-etched titanium surfaces can partially transform and grow upon soaking in NaOH. It can be confirmed using TEM/SAED

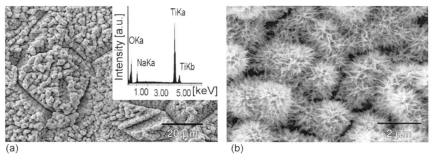

(a)   (b)

Figure 3.13 SEM micrograph of titanium after HCl etching at
50 °C/90 min +40 °C/60 min and subsequent alkali treatment
in NaOH at 60 °C for 24 h at different magnifications.
Inset: representative EDX analysis.

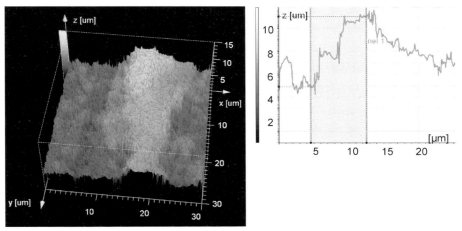

Figure 3.14 Surface profile of titanium after HCl etching at
50 °C/90 min +40 °C/60 min and subsequent alkali treatment
in NaOH at 60 °C for 24 h.

(Figure 3.16) that this modified layer is composed of nanocrystalline $Na_2Ti_3O_7$ and
$Na_2Ti_6O_{13}$ as well of some amorphous phase according to the following reactions:

$$3TiO_2 + 2NaOH \rightarrow Na_2Ti_3O_7 + H_2O \tag{3.4}$$

$$6TiO_2 + 2NaOH \rightarrow Na_2Ti_6O_{13} + H_2O \tag{3.5}$$

$$2TiH_2 + 2NaOH + 10H_2O \rightarrow Na_2Ti(OH)_4 + TiO_2 \cdot 9H_2O + H_2 \tag{3.6}$$

The bioactivity of surfaces thus prepared was tested *in vitro* using SBF (see Section 4.2).

The treatment of titanium by a two-step HCl and subsequent NaOH treatment
seems to be a suitable method for providing the titanium surface with bone-bonding

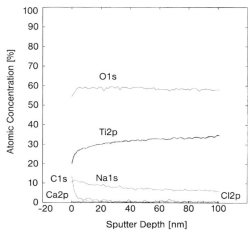

Figure 3.15 XPS analysis of titanium after HCl etching at 50 °C/ 90 min +40 °C/60 min and subsequent alkali treatment in NaOH at 60 °C for 24 h.

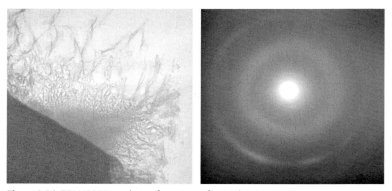

Figure 3.16 TEM/SAED analysis of titanium after HCl etching at 50 °C/90 min +40 °C/60 min and subsequent alkali treatment in NaOH at 60 °C for 24 h.

ability assumed from the ability to induce apatite formation from SBF. Acid etching of titanium in HCl under inert atmosphere leads to the formation of a uniform micro-roughened surface. The alkali modification shifted the hydrophobic state of the acid-etched surface to a hydrophilic one by creating a nano-substructure.

3.2.2
Micro Structured Titanium Surfaces Produced by Chemical and Electrochemical Etching

Ulrich Beck and Regina Lange

One of the drawbacks of corundum-blasted titanium surfaces is that the process leaves a substantial portion of the surface covered with residual corundum particles, which could lead to problems with medical implants [24]. These particles can be removed from the titanium surface using wet chemical etching.

Materials and Methods

After being corundum-blasted (see Section 3.1), cp-Ti specimens (grade 2, $d =$ 30 mm) were etched in an etching solution (37% HCl; 98% H_2SO_4; H_2O; volume ratio of $2 : 1 : 1$) for 5 min at 80 °C so as to remove the residual corundum particles from the titanium surface and achieve a leveled surface structure. Some of the specimens were also re-etched in the same etching solution with the aid of a cathodic electrical field ($U = 1.9$ V, $I = 1.1$ A) and were then rinsed in water.

Results

EDX investigations revealed that nearly all of the corundum particles were successfully removed from the surfaces, which were also leveled and showed a fine porous structure in the micrometer range and a mean roughness of approximately 1.5 μm. The ensuing cathode-aided etching process did not substantially modify either the surface structure or roughness (Figure 3.17). For information on surface characterization, see Part II.

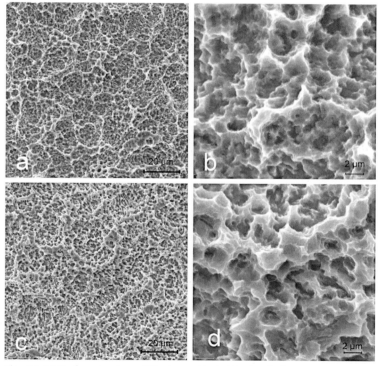

Figure 3.17 Surface structure of titanium subjected to chemical etching (a, b) and cathodic-supported chemical etching (c, d) (shown in two different magnifications, see scale bar).

3.2.3
Enhanced Hydrophobicity Achieved by Etching and Coating

Stefan Winter, Ulrich T. Seyfert, and Frank Aubertin

Surface Structuring

Materials for implants, particularly for coronary stents, must have specific properties, i.e. local biocompatibility and hemocompatibility, high strength under static or cyclic loading, cold deformability and radiopacity. Therefore niobium and tantalum, known as materials with a good biocompatibility [25] and radiopacity, were chosen as alloying elements for titanium-based alloys. The $(\alpha + \beta)$-alloys Ti10Nb, Ti20Nb, Ti10Ta, Ti20Ta and Ti30Ta were produced by arc melting. Cp-Ti grade 2, niobium and tantalum sheets of commercial purity were melted according to the description in Chapter 2. After homogenization in a vacuum at 1000 °C for 24 h the cigar-like rods were subjected to a thermomechanical treatment to produce a suitable microstructure for surface structuring. After a deformation of 60% above the β-transus temperature cylindrical platelets were machined from the cigars. A subsequent deformation of 75% in the $(\alpha+\beta)$-field at 650 °C and additional annealing produced a lamellar (Ti10Nb, Ti10Ta) and globular (Ti20Nb, Ti20Ta, Ti30Ta) microstructure, respectively. A schematic illustration of the thermomechanical treatment is given in Figure 3.18.

Because of the difference in the corrosion resistance of the α- and β-phase due to different amounts of the Nb and Ta alloying elements in both phases, a different etching behavior of the phases can be expected. In a first step an acidic corroding agent (aqueous solution of 0.2 mol L^{-1} HF and 2 mol L^{-1} HNO$_3$) at 20 °C with a variation of the etching time was tested for its ability to produce a defined topography and roughness at the surface of the samples. The results (arithmetic average height, R_a, and the maximum peak-to-valley height, PV) determined by means of white light interferometry are shown in Figure 3.19.

The R_a and PV values increase continuously with an increasing etching time from 2 to 10 min. After an etching treatment of about 10 min no significant dependence on

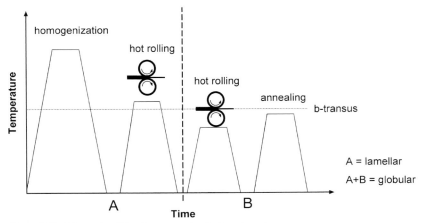

Figure 3.18 Schematic of the thermomechanical treatment.

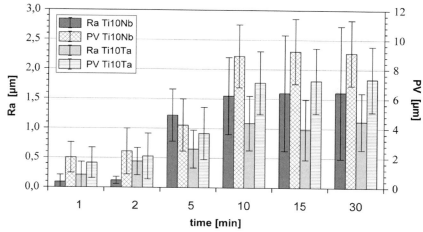

Figure 3.19 R_a and PV values of Ti10Nb and Ti10Ta.

the etching time can be detected. The high standard deviation values are an indication of an irregularly structured surface. Examples for the results of the etching procedure are given in Figures 3.20 and 3.21.

By this etching procedure, depending on the alloy used and the thermomechanical pre-treatment, three-dimensional images were produced of the lamellar and globular microstructure on the surface of the substrates.

Functional Electroconductive Coatings

The electrical properties and the wettability of the material surface play an important role in the interactions between an implant and blood. The electrical properties are responsible for interaction of the surface with proteins and it is reported that very highly hydrophobic surfaces suppress protein and platelet adsorption and thus ensure good blood compatibility [26,27].

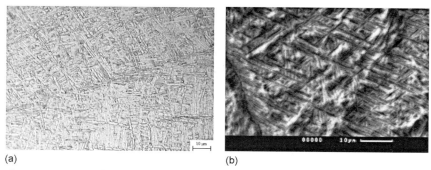

(a)   (b)

Figure 3.20 Ti10Nb after (a) the thermomechanical treatment and (b) a subsequent etching procedure; $R_a = 1.53\ \mu m$, PV $= 9.45\ \mu m$.

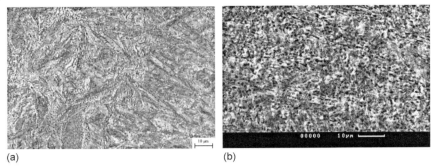

(a)   (b)

Figure 3.21 Ti20Nb after (a) the thermomechanical treatment and (b) a subsequent etching procedure; $R_a = 0.29 \, \mu m$, $PV = 2.56 \, \mu m$.

Therefore the aim of this investigation was the development of a superhydrophobic surface by a variation of the electrical properties and the chemical composition of the material surface. For this purpose metallographically polished cylindrical discs ($R_a \approx 10 \, nm$) of cp-Ti (see Chapter 2) were thermally oxidized, nitrided and carburized.

For this purpose cp-Ti discs were annealed in air, in a nitrogen atmosphere and in a CH_4 atmosphere, respectively, below the β-transus temperature of titanium (at 850 °C) with a variation of the annealing time in order to produce a layer thickness of approximately 100 nm. The thickness of the oxide, nitride and carbide layers was determined by means of ellipsometry (Section 4.1). The annealing parameters required to prepare an oxide layer of 100 nm thickness (70 min, 550 °C, air) were taken from [28]. The results of the nitriding and carburizing investigations are shown in Figure 3.22.

Figure 3.22 shows that the desired layer thickness of approximately 100 nm could be achieved by an annealing treatment for 4 h at 850 °C in N_2 and for 2 h at 750 °C in CH_4. XRD and Raman investigations showed that the prepared layers consist of an oxide layer (rutile), a nitride layer (TiN and TiN_2) and a carbide layer (TiC + amorphous and crystalline carbon).

For further investigations structured samples were selected to be thermally oxidized, nitrided and carburized. The samples chosen and their R_a values after the etching and coating procedures are shown in Table 3.1.

It is evident that neither the topography of the samples nor the surface roughnesses are significantly changed by the coating procedures.

Corrosion Properties

In a first step, corrosion tests were carried out on functionally coated samples of cp-Ti in physiological NaCl solution and current density–potential curves were recorded. The results are shown in Figure 3.23.

All titanium samples show a good corrosion resistance from 0 to 1000 mV whereas the CoCr alloy and the AISI 316L show a breakthrough potential of 600 and 300 mV. The polished and oxidized titanium samples possess very low current densities

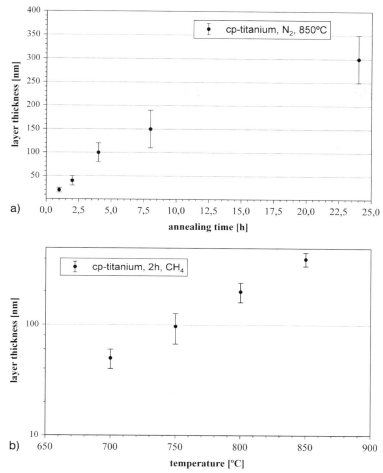

Figure 3.22 Resulting layer thickness of the nitriding (a) and carburizing (b) investigations on substrates of cp-Ti.

Table 3.1 Microstructure and R_a values of all materials structured by etching and differently coated.

Material	Structure	Polished, R_a (μm)	Etched, R_a (μm)	Etched, oxidized, R_a (μm)	Etched, nitrided, R_a (μm)	Etched, carburized, R_a (μm)
cp-Ti	polished	0.01	–	–	–	–
Ti10Nb	lamellar	0.02	0.33	0.35	0.44	0.34
Ti20Nb	globular	0.02	0.29	0.28	0.27	0.41
Ti10Ta	lamellar	0.02	0.68	0.77	0.81	0.69
Ti20Ta	globular	0.05	0.26	0.33	0.28	0.34
Ti30Ta	globular	0.02	0.34	0.29	0.42	0.31

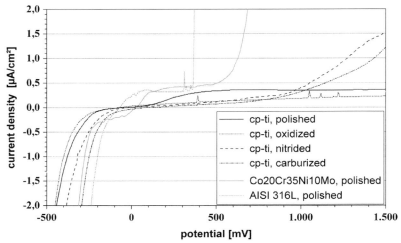

Figure 3.23 Current density–potential curves of surface-modified
cp-Ti and Co20Cr35Ni10Mo and AISI 316L for comparison.

throughout the passive range. The nitrided and carburized samples exhibit an
increase in their current densities at approximately 750 mV, which can be explained
by the beginning of an (anodic) oxidation process because TiN and TiC possess lower
enthalpies of formation than TiO_2 (rutile) [30].

The structured and coated TiNb and TiTa samples were also tested as regards their
corrosion properties. For purposes of clarity only the current densities at 400 mV are
given in Table 3.2.

The results correspond to those of the coated titanium samples. All materials tested
showed a very small current density in the passive potential range from 0 to 1000 mV.
Due to their two-phase microstructure the titanium alloys tested in their metallic state
show a slightly higher current density than cp-Ti.

Table 3.2 Passive current densities of the variously etched and
coated TiNb and TiTa alloys tested at 400 mV.

	Passive current density at 400 mV ($\mu A\,cm^{-2}$)			
Material	**Metallic**	**Etched + oxidized**	**Etched + nitrided**	**Etched + carburized**
cp-Ti	0.31	0.13	0.12	0.13
Ti10Nb	0.64	0.34	0.08	0.42
Ti20Nb	0.68	0.31	0.10	0.62
Ti10Ta	0.33	0.25	0.07	0.00
Ti20Ta	0.64	0.06	0.64	0.00
Ti30Ta	0.39	0.25	0.02	0.70

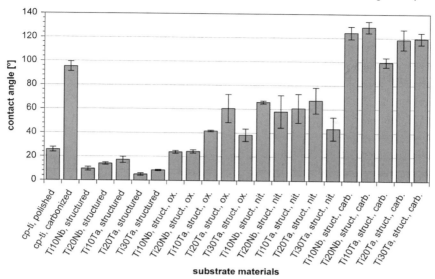

Figure 3.24 Contact angles of bi-distilled water on the differently structured and coated materials.

Wettability

The contact angles of bi-distilled water on the various coatings were detected by means of the sessile drop method. The results are shown in Figure 3.24.

It is shown that by means of the functional coatings the wettability of the etched TiNb and TiTa alloys could be reduced. By structuring alone the wettability was improved, which was not the desired aim, i.e. a low wettability. In contrast, the combination of structuring by etching and by functional coatings caused the contact angle of water to be increased up to 130° on the structured and carburized samples.

Hemocompatibility

For hemocompatibility tests spherical powders are required. These were produced using the rotating electrode process (REP). Titanium of commercial purity and the alloys described above were used as electrodes in this procedure [30]. Between the rotating electrode (12 000 rpm) and the water-cooled counter electrode (tungsten) an arc is ignited in a helium atmosphere. By this process spherical powders were produced and the desired powder grain size of 315–400 μm was obtained by sieving. The powders had also been oxidized, nitrided and carburized under the above conditions.

The hemocompatibility of the REP powders was tested *in vitro* in direct contact with human whole blood. The blood was mixed with citrate to avoid spontaneous clotting. Two hemo-rheological conditions were simulated in a centrifugal system at 37 °C, i.e. low shear stress (platelet-rich plasma, 150 g) and high shear stress (platelet-poor

Table 3.3 Nondimensional score system in comparison with the international standard ISO-10993-4.

Score system	ISO 10993-4 categories
Contact activation	Coagulation
Complement activation	Immunology
Thrombin generation	Thrombosis, coagulation
Fibrinogen–fibrin conversion	Thrombosis
Fibrinolysis	Thrombosis, coagulation
Hemolysis	Hematology
Proteolysis/burst	Immunology
Platelet activation	Platelets

plasma, 3000 g), simulating venous and arterial conditions. Various qualities of hemocompatibility were transformed to a nondimensional score-system, whereby 0 points is the best (hemocompatible) and 65 points the poorest (thrombogenious) result. Table 3.3 shows the comparability of this test method with the international standard ISO-10993-4. The results of the hemocompatibility tests are shown in Figure 3.25.

Since a score of 65 points represents absolute thrombogenicity, these results attest the good hemocompatibility of all materials tested. The maximum scores achieved were only 10 (Ti30Ta oxidized, low shear stress) and 17 (cp-Ti oxidized, high shear stress). The best results were achieved by nitrided Ti20Nb (2 points, low shear stress) and oxidized Ti10Ta (5 points, high shear stress).

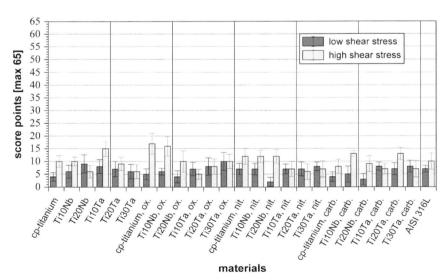

Figure 3.25 Results of the hemocompatibility tests of titanium alloys and AISI 316l for comparison.

3.3
Surface Modifications with Laser

André Gorbunoff, Wolfgang Pompe, and Michael Gelinsky

3.3.1
Introduction

The goal of surface processing in the biomaterials domain is to influence the behavior of the biological molecules and organisms in contact with these biomaterials in the desired way. Generally, there exist two significantly different basic approaches to surface lateral patterning with energy beams. *The mask projection* techniques use an energy or a particle beam to backlight a mask, with the mask image transferred to the surface by appropriate optics. This approach allows a rapid production of features over a large surface area, and is well suited to a high-volume production of a fixed pattern or shape. Since a relatively large surface area is exposed simultaneously, a high-energy source is needed for the mask projection. In contrast, *the direct write* techniques focus the entire beam of a much weaker source on the work surface and translate it to generate a desired pattern without a photomask.

A typical example of the mask projection technique is the resist-based silicon planar technology. In this technology, complex circuits are built on the surface of a silicon wafer in a succession of process steps. Each step contains a resist coating, an exposure, a development, an etching, a thin film deposition/implantation/etching and a resist removal.

The key role in the definition of the resulting feature size belongs to the exposure technique that transfers the desired pattern onto the photoresist film. For more than 30 years, optical projection lithography (photolithography) has been the workhorse of the semiconductor industry. By using a short-wavelength light source and advanced optical concepts one can achieve a resolution of down to 100 nm. This imposes, however, extremely stringent conditions on the wafer finish, the positioning tolerance of the wafer in the image plane, the resist thickness and the quality of the exposure system and the mask itself. These factors are significant driving forces towards elimination of the mask from microfabrication processes.

3.3.2
Laser Direct Writing

It can be shown that the laser direct write technique, despite the loss in throughput, has a number of considerable advantages as compared to the traditional projecting printing techniques. It has no problems associated with fabricating, inspecting, maintaining and cataloging masks, and with the yield loss due to any mask errors. No large field projection optics is needed. At present, laser direct writing has become a well-established technique in many micro- and macro-patterning applications.

In principle all technologically significant laser–solid interactions can be subdivided into physical (with no chemical transformations of the solid) and chemical ones.

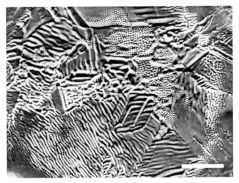

Figure 3.26 Periodical relief on the surface of polycrystalline gold after the action of 100 pulses of a Q-switched laser with an absorbed power density of 120 MW cm^{-2}. The scale bar is 100 μm. In some areas the relief mirrors the crystalline orientation of grains.

The physical laser–solid interactions are based primarily on the repeated heating and cooling the surface with the absorbed laser light. It can be accompanied by local melting, evaporation and formation of allotropic phases, which are widely used for welding, patterning and hardening of metals. Of importance for biomaterials applications is an often observed development of surface corrugations (Figure 3.26), which can result in the modification of the cells interaction with the laser-processed surfaces.

A number of mechanisms have been proposed to explain the formation of these surface reliefs:

- rapid thermal expansion or microstructural changes of the material in the region of laser influence, resulting in splitting of the surface layer of the target;
- momentum transferred to the liquid upon melting due to differences in densities between the solid and the liquid;
- recoil pressure pulse on the molten surface generated by the vaporized species;
- thermo-capillary instability in a melt heated non-uniformly from the surface inwards [31];
- explosive decay of the metastable liquid due to sub-surface superheating during the intensive vaporization by a surface heat source [32]; and
- electrohydrodynamic instability of a liquid metal in the electric field of the erosion plasma [33].

Most laser chemical processing technologies are based on the chemical interaction of the irradiated solid with surrounding reactive atmosphere. Consequently, by scanning a laser beam, it is possible to directly write a pattern on the surface of a substrate. This technique has been applied as a one-step manufacturing process for photomask fabrication and repair, writing electrically and optically active structural elements in prototype integrated circuits, manufacturing of customized devices and the information storage [34–36].

In *pyrolytic laser direct writing*, the laser beam focused on an absorbing substrate serves as a localized heat source that activates the desired chemical reaction on a given site of the surface. The resulting size of the reaction's spreading zone is determined by a complex conjunction of a variety of chemical and physical properties of the irradiated solid and the reaction products, as well as of the processes developing on the irradiated surface. Many of these processes depend nonlinearly on the laser intensity q. These intrinsic nonlinearities and feedbacks are known to allow fabrication of submicrometer structures below the conventional optical resolution limit [34,37–39]. In contrast, the conventional *photolytic* processes (like photography) are nearly linear, i.e. the developed thickness is only a function of the cumulative exposure independent of the intensity.

In laser chemical processing one considers commonly positive feedbacks when the reaction superlinearly accelerates with q due to increasing of the absorption as a result of

- surface defect generation;
- chemical reaction or phase transition;
- temperature dependence of optical, thermal or chemical constants;
- nonlinear mechanisms like multiphoton absorption, photo-induced carrier generation, photorefractive effects or resonant absorption on the laser-generated surface ripples.

The incorporation of these effects into the image-forming process results in a higher degree of reaction localization. Consider an oxidation reaction with an activation-type dependence of the reaction rate on temperature T:

$$w = w_0 \exp\left(-\frac{E_{act}}{R_g T}\right) \tag{3.7}$$

where E_{act} is the activation energy, and R_g the gas constant. Let a beam with a Gaussian radial distribution of beam intensity

$$q(r) = q_0 \exp\left(-\frac{2r^2}{R^2}\right) \tag{3.8}$$

be absorbed on a metal surface. This absorbed radiation will cause heating of the surface layer of the workpiece. Provided the heat losses are controlled only by the heat transport into the half-space of the specimen, after a time of few thermal diffusion time constants $\tau_{th} = R^2/\chi_s$ (where $\chi_s = k_s/c_s\rho_s$ is the heat diffusivity of the specimen material), the temperature distribution across the surface will reach the steady-state value [40]:

$$T(r) = T(0)\exp\left(-\frac{r^2}{R^2}\right) I_0\left(\frac{r^2}{R^2}\right) \tag{3.9a}$$

where I_0 is the modified Bessel function of the zero order,

$$T(0) = \frac{AP}{2\sqrt{\pi}k_s R} \tag{3.9b}$$

is the maximum temperature on the axis of the beam and AP is the total absorbed laser power.

Consider a time scale much greater than τ_{th}. Let us also assume that the main factor limiting the chemical reaction is the oxygen diffusion through the scale. This leads to the parabolic oxidation law

$$\frac{\partial h_o}{\partial t} = \frac{w}{h_{ox}} \tag{3.10}$$

typical for the high-temperature oxidation of metals [41,42], where h_{ox} is the oxide thickness.

Model calculation of the laser oxide layer thickness and the optical transmission distribution of the oxidation reaction zone in a thin titanium film on a quartz glass substrate in oxidizing atmosphere based on the system Eqs. (1) to (4) are presented in Figure 3.27. (The use of a glass substrate reduces χ_s and increases $T(0)$.)

The writing process with a strong positive feedback becomes rather unstable and very sensitive to the laser power instabilities [39,43]. Gorbunov *et al.* [44,45] have demonstrated the advantage of the system with an absorptivity A which *decreases* with the development of the chemical reaction. They considered a system of thin absorbing film on a transparent substrate, in which the laser-induced reaction slows down just after beginning due to the fading of the absorbing film upon through-oxidation. This helps to stabilize the reaction and prevents it from broadening laterally. Similar negative feedbacks can be achieved by decreasing the absorptivity due to bleaching [46] or a phase transition [47], or increasing heat conductivity with T. For example, the jump of the reflectivity of silicon at the melting point enabled Müllenborn et al. [47] to form 70 nm wide trenches in silicon through direct laser etching in chlorine ambient.

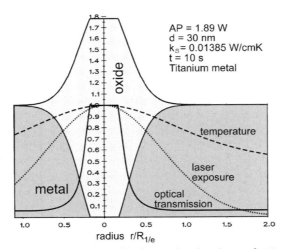

Figure 3.27 Modeling of the laser-induced oxidation of a 30 nm thick titanium film on glass. $R_{1/e} = 0.756$ mm, the absorbed laser power $AP = 1.89$ W, the interaction time $t = 10$ s $\gg \tau_{th} \approx 1$ s, E_{act} and w_0 were taken to be 280 kJ mol^{-1} and 330 m^2 s^{-1} [49].

3.3.3
Surface Microprocessing of Titanium Films by Self-limited Oxidation

Gorbunov *et al.* [44,45] have experimented with thin (3 to 60 nm) titanium films deposited on glass substrates. The patterning of the samples was performed in air by a commercial laser scanning microscope (LSM). The radiation of an argon ion laser operating at two wavelengths $\lambda = 488$ and 514 nm was focused by a $20 \times /0.5$ or a $50 \times /0.85$ objective to a minimal spot diameter on the surface of the sample of $D_{sp} = 2R_{sp} = 890$ nm or 500 nm (at the intensity level $1/e^2$), respectively. The poor heat conductivity of the glass substrates permits one to reduce considerably the working laser power P (measured on the surface of the sample) to several milliwatts. With an internal deflection system, the radiation was scanned line-by-line across the sample with a rate V of 0.5 to 110 mm s^{-1}. The same microscope with a $\times 100$ attenuation was used for optical imaging of the samples in both reflection and transmission modes. The height and width of the written patterns were investigated using an atomic force microscope (AFM).

The experiments have shown that the scanning under optimal process parameters resulted in the formation of line gratings well contrasting optically with the initial titanium film. Figure 3.28 shows an optical transmission image of an oxide line grating formed in a 15 nm thick titanium film.

The fact that the oxidation really took place under the presented experimental conditions was verified first by performing the experiments under nitrogen blowing (no patterning took place) and second by energy dispersive X-ray spectroscopy (EDXS). To minimize the estimated X-ray excitation depth to about 100 nm, the EDXS analysis was carried out with an accelerating voltage of 5 kV. An enhanced concentration of oxygen and a decreased signal of silicon (an element of the substrate) were revealed in the processed regions (Figure 3.29) indicating thickening and

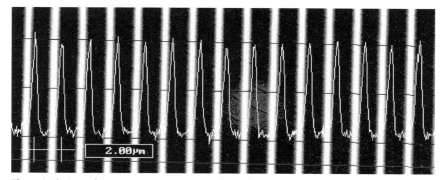

Figure 3.28 Optical transmission image (vertical lines) and the transmission profile across the lines of a laser-written oxide grating on a 15 nm titanium film on glass [45]. The actual line width might be broadened by the diffraction. The transmission profile is somewhat shifted to the right against the original for clarity. The lines with a period of 2 μm were written with $D_{sp} = 890$ nm, $P = 10.5$ mW and $V = 1.1$ mm s^{-1}.

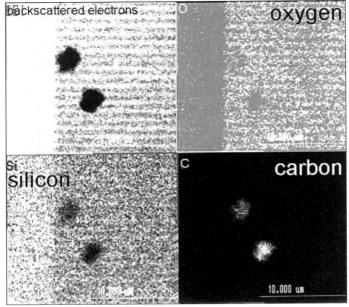

Figure 3.29 EDXS element mapping of a 60 nm titanium film on glass laser-patterned at $P = 10.5$ mW and $V = 1.1$ mm s^{-1} using a 20 × / 0.5 objective [45]. The laser-written lines on the right of the upper left image (backscattered electrons) show an increased presence of oxygen (upper right image) and a reduced background of silicon (lower left image) from the substrate due to the additional thickness of the oxide lines. Two dark blobs on the backscattered electron image are carbon-containing contamination (lower right). The brighter the image the higher the content of the element.

oxidation of the titanium film on the silicon oxide-containing glass substrate upon illumination. A typical AFM image of one of the test patterns and the height profile across it is presented in Figure 3.30.

A possible scenario of the writing process is shown in Figure 3.31. When reaching a non-exposed portion of the titanium film surface, the laser beam will be absorbed according to the initial absorptivity A, and the temperature starts to increase. Since the characteristic exposure time $D_{sp}/V \geq 4.5\,\mu s \gg \tau_{th} \approx 100$ ns (with χ_s of glass), one can assume that the temperature (at least near the critical process parameters) is close to the stationary value during the whole process (Eq. 3.9). $T(0)$ is thus determined only by the instantaneous laser power absorbed AP. The laser power P remains always the same. In turn, the optical absorptivity A depends both on the optical constants of the film material and on the metal film thickness d.

The dependence $A(d)$ of a titanium film on glass calculated following the formalism of Born and Wolf [48] is presented in Figure 3.32. It predicts that A should steeply decrease from about 0.4 (the measured $A = 0.51$ for $\lambda = 488$ and 514 nm for films thicker than 30 nm) to 0 when d becomes smaller than 10 nm. The absorption of light in the growing oxide should be negligible because its absorption coefficient α (cm^{-1}) is factor of 10^2-10^4 smaller than that of titanium.

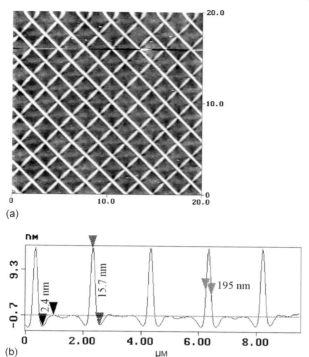

(a)

(b)

Figure 3.30 (a) AFM image and (b) the corresponding height profile of a typical laser-written line pattern [45]. $d = 15$ nm, $D = 500$ nm, $P = 6.75$ mW, $V = 1.1$ mm s^{-1}.

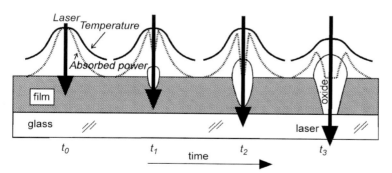

Figure 3.31 Mechanism of the self-limitation of laser-induced thermochemical reactions in thin absorbing films on a transparent substrate: (t_0) the beginning of the reaction with a uniform absorptivity; (t_1) the initial development of the reaction with decreasing of absorptivity on the axis; (t_2) the formation of an oxide window in the center of the reaction zone; (t_4) the lateral growth of the oxide window and the reduction of the reaction temperature.

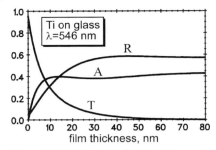

Figure 3.32 The optical reflectivity R, transmissivity T and absorptivity A of a titanium film on glass calculated after Born and Wolf [48] for $\lambda = 546$ nm and a complex refractive index $\tilde{n} = 2.53–i3.33$ [45].

In the first moments of time, the oxidation reaction is sharply localized at the top of the temperature profile. This leads to practically instantaneous (as compared to the characteristic scanning time scale D_{sp}/V) through-oxidation of the film and the formation of a small oxide window in the center of the reaction zone. The reaction will then run further and the oxide window will expand until the power of the reaction heat release cannot compensate the decrease in the absorbed laser power. After that the maximum temperature starts to fall. Since the activation energy of titanium oxidation E_{act} is fairly high (185 to 280 kJ mol^{-1} [41,42,49]), the reaction rate w in the low-temperature region is extremely sensitive to the reaction temperature T. For example, a 10% reduction of T at $T = 1400$ K (typical temperatures of the process estimated using Eq. (3.9) results in an order of magnitude deceleration of the reaction, which practically means its abrupt cut-off.

The laser-assisted oxidation of thin titanium films in air is a complex process which involves positive (the increase in the reaction rate and the reaction heat release with temperature) and negative (the enlightenment of the metal film upon oxidation) feedbacks. It allows one to fabricate novel nanometer-patterned titanium surfaces which are promising objects for biomaterial applications. The behavior of osteoblast-like cells on such structured model surfaces was investigated in collaboration and the results are presented in Section 3.2 of Part III.

3.4
Plasma Processes

Ulrich Beck and Regina Lange

Materials and Methods
All of the mechanical processes except for polishing allow for relatively simple and inexpensive titanium surface structuring in the 0.5 μm $\leq R_a \leq$ 6 μm roughness range (see Section 3.1). Plasma processes can be used for surface structuring outside of this range. Electrolytic micro-plasma processes (EMPs) can be used to create surfaces with relatively level structures (including fine porous oxidized surfaces) in the

$50\,\text{nm} \le R_a \le 1\,\mu\text{m}$ range. Vacuum plasma injection processes are suitable for the range $10\,\mu\text{m} \le R_a \le 50\,\mu\text{m}$.

Electrolytic Micro-plasma Process (EMP)   EMP (also known as micro-arc and micro-spark discharging) is realized at the phase boundary between solid and liquid electrodes. Solid electrodes are metals such as aluminum, tantalum, magnesium or (as in the present case) titanium. Liquid electrodes are realized using a suitable aqueous electrolyte. As is well known, plasma chemical reactions of this type involve a series of thermal and chemothermal processes. Owing to water electrolysis a thin layer of water vapor is formed on the titanium surface when a sufficiently high voltage is generated between the electrodes. Thus the ohmic resistance increases at the phase boundary. Following this, the aqueous electrolyte reaches the boiling point owing to the presence of Faraday current. The resulting arc discharge then generates high current density ionic plasmas.

Owing to the locally stochastic distribution of the ohmic resistance and electrical field strength on the titanium surface, the plasmas generated at the phase boundary are also locally distributed micro-plasmas. This also results in local melting of the titanium surface and structuring thereof. The surface coating, surface coating composition and surface structure can be modified by varying the electrical process and the composition of the electrolyte.

In one variant (method 1), a wet chemical etching process electrolyte (37% HCl; 98% H_2SO_4; H_2O in a 2:1:1 ratio; see Section 3.3) was used at $20\,^\circ$C, 100 V and for 1 min. In the second variant (method 2), a solution comprising 300 mL of 96% H_3PO_4, 500 mL of H_2O and 25 g of $CuSO_4$ was used as an electrolyte at $20\,^\circ$C and 150 V for 3 min.

Vacuum Plasma Injection   In our vacuum plasma injection process, the evacuated coating chamber is filled with argon up to a pressure of approximately $10^4\,$Pa. Following ignition, titanium powder is injected into the hot gas plasma (10 000 K) and is applied to the target substrate via the plasma spray at a velocity of approximately $1000\,\text{m}\,\text{s}^{-1}$, thus creating an oxide-free surface coating [50]. By varying the process parameters the layer structure can be modified to a substantial extent. When the particles are molten completely and impact the substrate at a high velocity, extremely dense layers are formed. In cases where only the surface of the powder particles is molten, an open-pored layer is formed. The structure of this layer is determined by particle size, the density of the melted-on surface layer and the impact velocity of the particles at the substrate. In most cases before the plasma spraying process a compact adhesive layer is formed first, and then the function layer with the desired target structure is applied on this layer.

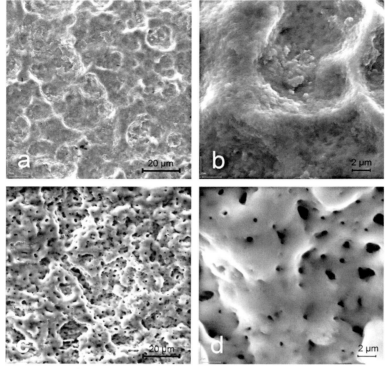

Figure 3.33 Surface structures of micro-plasma-treated titanium specimens (EMP), shown at two different magnifications: (a, b) method 1, (c, d) method 2.

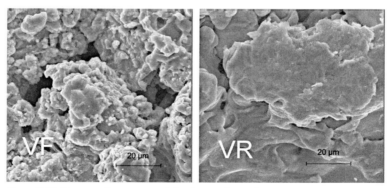

Figure 3.34 Vacuum plasma sprayed titanium surfaces: fine vacuum plasma (VF) and rough vacuum plasma (VR) procedures.

Results

Results with Regard to EMP   The micro-plasma process leveled the surfaces and resulted in the formation of a surface coating of varying properties. These surfaces may also have ceramic properties depending on the composition of the electrolyte used (see Figure 3.33).

Results with Regard to Vacuum Plasma Injection   Highly structured surfaces with a mean roughness value of $R_a = 21\,\mu m$ (small-grained titanium powder → fine vacuum plasma injection) and $R_a = 48\,\mu m$ (large-grained titanium powder → rough vacuum plasma injection) were generated (Figure 3.34). It should be noted that despite the lower degree of roughness, the fine vacuum plasma injection procedure resulted in a higher porosity of the surface than with the rough vacuum plasma injection procedure. The surface generated by the latter process appeared to be smoother and less structured. For more information about surface properties see Part II.

References

1 Lange, R., Lüthen, F., Beck, U., Rychly, J., Baumann, A. and Nebe, B. (2002) Cell–axtracellular matrix interaction and physico-chemical characteristics of titanium surface depend on the roughness of the material. *Biomol. Eng.*, **19**, 255–261.

2 Linez-Batailon, P., Monchau, F., Bigerelle, M. and Hildebrand, H.F. (2002) *In vitro* MC3T3 osteoblast adhesion with respect to surface roughness of Ti6Al4V substrates. *Biomol. Eng.*, **19**, 133–141.

3 Kirbs, A., Kange, R., Nebe, B., Rychly, R., Baumann, A., Neumann, H.-G. and Beck, U. (2003) Methods for physical and chemical characterisation of surface of titanium implants. *Mater. Sci. Eng. C*, **23**, 425–429.

4 Jayaraman, M., Meyer, U., Bühner, M., Joos, U. and Wiesmann, H.-P. (2004) Influence of titanium surfaces on attachment of osteoblast-like cells *in vitro*. *Biomaterials*, **25**, 625–631.

5 Bigerelle, M. and Anselme, K. (2005) A kinetic appraoch to osteoblast adhesion on biomaterial surface. *J. Biomed. Mater. Res.*, **75A**, 530–540.

6 Huang, H.-H., Ho, C.-T., Lee, T.-H., Lee, T.-L., Liao, K.-K. and Chen, F.-L. (2004) Effect of surface roughness of ground titanium on initial cell adhesion. *Biomol. Eng.*, **21**, 93–97.

7 Eisenbarth, E., Linez, P., Biehl, V., Velten, D., Breme, J. and Hildebrand, H.F. (2002) Cell orientation and cytoskeleton organisation on ground titanium surfaces. *Biomol. Eng.*, **19**, 233–237.

8 Lincks, J., Boyan, B.D., Blanchard, C.R., Lohmann, C.H., Liu, Y., Cochran, D.L., Dean, D.D. and Schwartz, Z. (1998) Response of MG63 osteoblast-like cells to titanium and titanium alloy is dependent on surface roughness and composition. *Biomaterials*, **19**, 2219–2232.

9 Schuh, A., Holzwarth, U., Kachler, W., Göske, J. and Zeiler, G. (2004) Oberflächenuntersuchungen an Al2O3-raugestrahlten Titanimplantaten in der Hüftendoprothetik. *Der Orthopäde*, **8**, 905–910.

10 Albrektson, T., Brenemark, P.-I., Hansson, H.-A., Ivarsson, B. and Johsson, U. 1982, Ultrastructural

analysis of the interface zone of titanium and gold implants, in *Clinical Applications of Biomaterials* (eds A.J.C. Lee,T. Albrektsson,P-.I. Brenemark),Wiley,Chichster, pp. 167–177.

11 Szmukler-Moncler, S., Testori, T. and Bernard, J.P. (2004) Etched implants: a comparative study analysis of four implant systems. *J. Biomed. Mater Res. B: Appl. Biomater.*, **69B**, 46–57.

12 Kieswetter, K., Schwarz, Z., Dean, D.D. and Boyan, B.D. (1996) The role of implant surface characteristics in the healing of bone. *Crit. Rev. Oral Biol. Med.*, **7** (4), 329–345.

13 Rupp, F., Schneideler, L., Rehbein, D., Axmann, D. and Geis-Gerstofer, J. (2004) Roughness induced dynamic changes of wettability of acid etched titanium implants. *Biomaterials*, **25**, 1429–1438.

14 Lampin, M., Warocquier, C., Legris, C., Degrange, M. and Sigot-Luizard, M.F. (1997) Correlation between substratum roughness and wettability, cell adhesion, and cell migration. *J. Biomed. Mater. Res.*, **36**, 99–108.

15 Deligianni, D.D., Katsala, N., Ladas, S., Sotiropoulou, D., Amedee, J. and Missirlis, Y.F. (2001) Effect of surface roughness of the titanium alloy Ti–6Al–4V on human bone marrow cell response and on protein adsorption. *Biomaterials*, **22**, 1241–1251.

16 Rupp, F., Schneideler, L., Olshanska, N., de Wild, M., Wieland, M. and Geis-Gerstorfer, J. (2006) Enhancing surface free energy and hydrophilicity through chemical modification of micro-structured titanium implant surface. *J. Biomed. Mater. Res.*, **76A**, 323–334.

17 Conforto, E., Caillard, D., Aronsson, B.-O. and Descouts, P. (2004) Crystallographic properties and mechanical behaviour of titanium hydride layer grown on titanium implants. *Philos. Mag.*, **84** (7), 631–645.

18 Lisowski, W., van den Berg, A.H.J. and Smithers, M. (1998) Characteritzation of titanium hydride film after long-term air interaction: SEM, ARXPS and AES depth profile studies. *Surf. Interf. Anal.*, **26**, 213–219.

19 Pouilleau, J., Devilliers, D., Garrido, F., Durand-Vidal, S. and Mahé, E. (1997) Structure and composition of passive titanium oxide films. *Mater. Sci. Eng.*, **B47**, 235–243.

20 Li, P., Ohtsuki, C., Kokubo, T., Nakanishi, K., Soga, N., Nakamura, T., Yamamuro, T. and de Groot, K. (1994) The role of hydrated silica, titania, and alumina in inducing apatite on implants. *J. Biomed. Mater. Res.*, **28**, 7–15.

21 Li, P., Kangasniemi, I., de Groot, K. and Kokubo, T. (1994) Bonelike hydroxy-apatite induction by a gel-derived titania on a titanium substrate. *J. Am. Ceram. Soc.*, **5**, 1307–1312.

22 Jonášov,á L., Müller, F., Helebrant, A., Strnad, J. and Greil, P. (2002) Hydroxyapatite formation on alkali-treated titanium with different content of Na $^+$ in the surface layer. *Biomaterials*, **23**, 3095–3101.

23 Jonášová L., Müller, F., Helebrant, A., Strnad, J. and Greil, P. (2004) Biomimetic apatite formation on chemically treated titanium. *Biomaterials*, **25**, 1187–1194.

24 Schuh, A., Holzwarth, U., Kachler, W., Göske, J. and Zeiler, G. (2004) Ober-flächenuntersuchungen an Al_2O_3–raugestrahlten Titanimplantaten in der Hüftendoprothetik. *Der Orthopäde*, **33**, 905–910.

25 Helsen, J.A. and Breme, J.A. (1998) *Metals as Biomaterials*, John Wiley, New York.

26 Lin, Y.S., Hlady, V. and Janatova, J. (1992) Adsorption of complement proteins on surfaces with a hydro-phobicity gradient. *Biomaterials*, **13**, 497–504.

27 Sun, T., Tan, H., Han, D., Fu, Q. and Jiang, L. (2005) No platelet can adhere-

largely improved blood compatibility on nanostructured superhydrophobic surfaces. *Small*, **10**, 959–963.

28 Velten, D. (2000)Herstellung von Titandioxid-Schichten auf Titanwerkstoffen mit Hilfe der thermischen und anodischen Oxidation, sowie durch den Sol-Gel-Prozeß und deren Charakterisierung, Diplomarbeit, Lehrstuhl für Metallische Werkstoffe, Universität des Saarlandes, Saarbrücken.

29 Kubaschewski, O., Alcock, C.B. and Spencer, P.J. (1993) *Materials Thermochemistry*, 6th edn, Pergamon Press, Oxford.

30 Biehl, V., Wack, T., Winter, S., Seyfert, U.T. and Breme, J. (2002) Evaluation of the haemocompatibility of titanium based biomaterials. *Biomol. Eng.*, **19**, 97–101.

31 Levchenko, E.B. and Chernyakov, A.L. (1983) Instability of capillary waves in non-uniformly heated melt under the action of laser radiation. *Fiz. Khim. Obrabotki Mater.*, **1**, 129–130 (in Russian).

32 Mazhukin, V.I. and Samarskii, A.A. (1994) Mathematical modelling in the technology of laser treatments of materials. *Surv. Math. Ind.*, **4**, 85–149.

33 Vladimirov, V.V. and Golovinskij, P.M. (1982) Excitation of capillary waves on the liquid metal surface that contacts with unstable plasma. *Pisma Zh. Eksp. Teor. Fiz.*, **81**, 1464–1469. (in Russian).

34 Bäuerle, D. (1996) *Laser Processing and Chemistry*, Springer-Verlag, Berlin/ Heidelberg.

35 Madou, M. (1997) *Fundamentals of Microfabrication*, CRC Press, Boca Raton, FL.

36 Metev, S.M. and Veiko, V.P. (1998) *Laser-Assisted Microtechnology*, Springer-Verlag, Berlin/Heidelberg.

37 Kirichenko, N.A. and Luk'yanchuk, B.S. (1983) Laser activation of oxidizing reactions at the metal surface. *Kvantovaya Electron.*, **10**, 819–825 (in Russian).

38 Liu, Y.S. (1989) Laser Microfabrication. in *Thin Film Process and Lithography* (eds D.J. Ehrlich and J.Y. Tsao), Academic Press, London, pp. 3–84.

39 Arnold, N. (1995) On the spatial confinement in energy beam microprocessing. *J. Appl. Phys.*, **78**, 4805–4807.

40 von Allmen, M. (1987) *Laser Beam Interactions with Materials*, Springer-Verlag, Berlin.

41 Bai, A.S., Lainer, D.I., Slesareva, E.N. and Tsypin, M.I. (1970) *Oxidation of titanium and its alloys*, Metallurgija, Moscow. (in Russian).

42 Kofstad, P. (1988) *High Temperature Corrosion*, Elsevier, London.

43 Ehrlich, D.J. and Tsao, J.Y. (1984) Nonreciprocal laser-microchemical processing: spatial resolution limits and demonstration of 0.2 μm linewidth. *Appl. Phys. Lett.*, **44**, 267–269.

44 Gorbunov, A.A., Eichler, H., Pompe, W. and Huey, B. (1996) Lateral self-limitation in the laser-induced oxidation of ultrathin metal films. *Appl. Phys. Lett.*, **69**, 2816–2818.

45 Gorbunov, A.A., Pompe, W., Eichler, H., Huey, B. and Bonnell, D.A. (1997) Nanostructuring of laser-deposited Ti films by self-limited oxidation. *J. Am. Ceram. Soc.*, **80**, 1663–1667, 2468.

46 Gochiyaev, V.Z., Korol'kov, V.P., Sokolov, A.P. and Chernukhin, V.P. (1989) Half-tone optical recording on a-Si films. *Kvantovaja Electron.*, **16**, 2343–2348 (in Russian).

47 Müllenborn, M., Dirac, H. and Petersen, J.W. (1995) Silicon nanostructures produced by laser direct writing. *Appl. Phys. Lett.*, **66**, 3001–3003.

48 Born, M. and Wolf, E. (1993) *Principles of Optics*, Pergamon Press, London.

49 Prokhorov, A.M., Konov, V.I., Ursu, I. and Mihailescu, I.N. (1988) *Interaction of Laser Radiation with Metals*, Nauka, Moscow (in Russian). [A concise

English translation: I. Ursu, A. M. Prokhorov, I. N. Mihailescu, V. I. Konov, *Laser Heating of Metals*, Adam Hilger Series on Optics and Optoelectronics, Institute of Physics Publishing, Philadelphia, 1990].

50 Gruner, H. (2001)Thermal spray coatings in medical applications, in *Titanium in Medicine.* (eds. D.M. Brunette, P. Tengvall, M. Textor and P. Thomsen), Springer, Berlin, pp. 388–404.

4
Chemical Surface Modifications

4.1
Coatings with Passivating Properties

4.1.1
Sol–Gel Coatings

Stefan Winter, Dirk Velten, Frank Aubertin, Bettina Hoffmann, Frank Heidenau, Günter Ziegler

Materials and Methods

Cylindrical specimens of the β- and near-β-titanium alloys Ti13Nb13Zr, Ti15Mo5Z-r3Al, Ti30Nb and Ti30Ta, which were prepared in the martensitic state (groove rolling at 850 °C with subsequent water quenching), and rods of commercial cp-titanium (α), Ti6Al7Nb (α + β) and Ti6Al4V (α + β) with a composition according to DIN ISO 5832 were used. Discs (diameter of 15 mm, 2 mm thick) were turned off and mechanically abraded with SiC paper. After polishing with a SiO_2 suspension to a mirror-like surface ($R_a < 10$ nm) the samples were ultrasonically rinsed in ethanol and dried in a hot air flow. The surface quality of the samples was inspected by means of light optical microscopy in polarization mode.

Sol–Gel Procedure

In order to prepare the starting sols for coatings consisting of TiO_2, ZrO_2, Nb_2O_5 and Ta_2O_5, respectively, mixtures of a metalorganic precursor (niobium(V) ethoxide, tantalum(V) ethoxide, titanium(IV) butoxide, zirconium(IV) butoxide), a solvent (butanol) and a chelating agent (acetylacetone) in the molar ratio $1 : 20 : 1$ were prepared. The solution was stirred for 24 h. The polished discs were spin coated (6000 rpm, 60 s) with the sols and then dried for 30 min at 150 °C. By repeating the cycles of coating and subsequent drying, thicker oxide layers can be achieved. After drying, the samples were heated to 400 °C in air at a rate of 4 K min^{-1} and tempered at this temperature for 1 h to remove the organic residuals. Finally, the oxide layers were annealed at temperatures between 450 and 700 °C. The preparation of the sol–gel oxide layers, e.g. TiO_2, is summarized in Figure 4.1.

Metallic Biomaterial Interfaces. Edited by J. Breme, C. J. Kirkpatrick, and R. Thull
Copyright © 2008 WILEY-VCH Verlag GmbH & Co. KGaA, Weinheim
ISBN: 978-3-527-31860-5

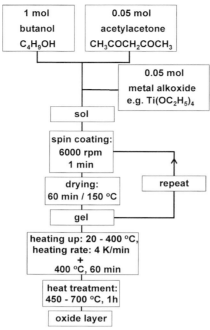

Figure 4.1 Schematic description of the preparation of oxide layers by means of the sol–gel process.

The parameters required to produce organic-free oxide layers were determined by means of differential thermal analysis, thermogravimetry (DTA-TG) and CHN analyses. The powder samples, which were necessary for these analyses, were prepared by adding water to the sol. The resulting hydrolysis–condensation reaction lead to the precipitation of the oxide which was cleansed and dried at 150 °C until the solvent had evaporated and the powder remained.

Characterization Techniques

DTA-TG measurements were carried out with the dried oxide powder in a synthetic air flow at a heating rate of $1\,\mathrm{K\,min^{-1}}$. Alumina was used as a reference for DTA. Potential organic residues in the heat treated oxide were detected by their possible carbon content by means of CHN analysis. The thicknesses of the oxide layers were determined ellipsometrically. Phase analyses were carried out by glancing-angle X-ray diffraction (XRD) (Cu K_α, $\Theta = 1°$). The patterns were identified by the use of the JCPDS. The morphology was observed using transmission electron microscopy (TEM). The grain size of the oxide crystallites was investigated by glancing-angle XRD using the method according to Scherrer [1]. The topography and the roughness of the surfaces were determined by means of atomic force microscopy (AFM) and high-resolution scanning electron microscopy (HRSEM). The sessile drop method was used to determine the contact angle of bi-distilled water on the oxide layers. The samples were cleansed in cyclohexane, acetone, ethanol and twice in distilled water to remove impurities and adjust reproducible surface qualities for the measurement of the contact angle. Current density–potential curves were used to determine the corrosion properties of the coated samples under physiological conditions (0.9% NaCl at 37 °C in H_2O, pH $= 7.4$). The potential between the sample (working electrode) and the reference electrode (Argenthal Ag/AgCl, $+207$ mV to SHE) was raised from -1 to $+3$ V with a scan rate of 0.1 mV s^{-1}.

Analysis of Organic Residues in the Sol–Gel Oxides

To prepare oxide layers with a high purity, the samples have to be heat treated at temperatures that are high enough to release all organic contamination. Information about the required temperatures can be obtained from the DTA-TG curves. As an example the DTA-TG curve of Ta_2O_5 is shown in Figure 4.2.

Because of the release of organic residuals from the gel, a weight loss in the TG curve occurs amounting to 18% and is completed at 430 °C. The DTA measurement shows a broad exothermic peak from 200 to 400 °C which is caused by the exothermic reactions due to the combustion of hydrogen and carbon from the organic constituent [2]. The small exothermic peak at 690 °C, which shows no correlating weight loss, can be assigned to the phase transition from amorphous to crystalline oxide. Because no further weight loss or exothermic peak appears above 430 °C, the removal of organic residuals is complete at this temperature. The behavior of the other oxides is comparable to that of Ta_2O_5. At a temperature of at least 450 °C the weight losses in the TG curves are at an end and the exothermic reaction which is caused by the combustion of the organic residues is also completed. A crystallization peak can also be observed with Nb_2O_5 at 555 °C, but not with TiO_2 and ZrO_2. The reason for the missing crystallization peaks in the DTA curves might be the overlap of the strong exothermic reaction of the combustion over the small peak of the crystallization of TiO_2 and ZrO_2 which takes place in the same temperature range from 200 to 400 °C.

In order to confirm the results of the DTA-TG measurements, the oxides were investigated by means of CHN analysis. The carbon content was detected from dried oxide powders which were heat treated at 300, 400 and 500 °C for 1 h. With increasing

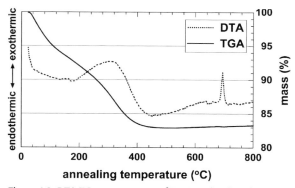

Figure 4.2 DTA-TG measurement of Ta_2O_5 sol–gel oxide.

Table 4.1 Carbon content (wt%) of the sol–gel oxides after various heat treatments.

	Nb_2O_5	Ta_2O_5	TiO_2	ZrO_2
150–400 °C, 4 K min^{-1} + 400 °C, 60 min	0.04	0.08	0.08	0.79
500 °C, 1 h	0.02	0.05	0.08	0.58

temperature the carbon content decreases from 1.39% at 300 °C to 0.29% at 400 °C and 0.11% at 500 °C. To determine the influence of the temperature control on the release of the organic residues, the dried oxides were heated from 150 to 400 °C at a rate of 4 K min^{-1} and tempered at 400 °C for 60 min similar to the parameters in the DTA-TG measurements. Additionally, this powder was heat treated at 500 °C for 1 h, simulating a heat treatment for the adjustment of the oxide properties. The results are summarized in Table 4.1.

After passing through the temperature ramp and the subsequent annealing (400 °C, 60 min), the oxides have a carbon content of 0.04–0.08% corresponding to the content of cp-titanium which according to the standards amounts to $\leq$0.08 wt%. This content is lower than that of the powders heat treated at 400 °C for 1 h without a temperature ramp (0.11–0.29%). The continuous heating results in a slow densification of the oxide. Thus the organic residuals can easily leave the oxide network which is still open. This is in contrast to the direct annealing at 400 °C where the network is densified very quickly, thus hindering the release of the organic compounds. The slightly higher carbon content of ZrO_2, as compared to that of the other sol–gel oxides, is caused by the stronger binding of the chelating agent acetylacetone to the zirconium central atom [3].

By means of DTA-TG measurements and CHN analyses, the optimized parameters of the heat treatment with regard to organic-free oxides can be defined. The slow heating to 400 °C combined with a subsequent annealing at 400 °C for 60 min results in an oxide which is almost free of organic residues.

Layer Thickness

Table 4.2 shows the thicknesses of the oxide layers on TiAl6Nb7 after drying and after annealing at 500 °C for 1 h measured by means of ellipsometry. The values for a single coating amount to 31 to 66 nm depending on the reactivity of the starting sol.

To prepare thicker layers the coating cycle can be repeated several times. Because of the release of organic residues and a simultaneously occurring densification, the thickness decreases after the annealing as compared to the dried condition. The observed shrinkage amounts to 24 to 33% and is higher for Nb_2O_5 and Ta_2O_5 as compared to TiO_2 and ZrO_2. The reason might be the higher reactivity of the niobium and tantalum precursors leading to a higher number of reacted precursor molecules in the hydrolysis–condensation reaction because molecules which already reacted during the drying procedure are not disposable during annealing. Thus, already during drying, the niobium and tantalum gels release more water and alcohol,

Table 4.2 Thickness of the oxide layer per coating cycle (spin-coating).

	Nb_2O_5	Ta_2O_5	TiO_2	ZrO_2
After drying (nm)	53	86	48	46
After annealing (nm)	40	66	34	31
Shrinkage (%)	25	24	30	33

resulting in fewer released molecules during the annealing and therefore in a lower shrinkage. Coating experiments with the other substrate materials show that there is no influence of the substrate on the thickness of the oxide layers in the dried condition.

Phase Analysis

The glancing-angle XRD measurements were carried out on substrates of the β- and near-β-titanium alloys coated with the different oxides and heat treated at temperatures between 400 and 700 °C. To avoid negative influences on the mechanical properties of the substrate metals (e.g. by means of uncontrolled grain growth), no heat treatments were carried out at a temperature higher than 700 °C. The thickness of the layers amounts to 121 to 131 nm.

The diffraction pattern for the Nb_2O_5 samples (substrate material: Ti30Ta) annealed at 400 and 500 °C shows peaks only from the substrate, indicating an amorphous structure of the oxide film. Starting with an annealing at 550 °C, X-ray reflections of crystalline oxide, monoclinic Nb_2O_5, are observed. The phase transition amorphous/crystalline is observed at a temperature between 500 and 550 °C, which is lower than the transition temperature of Nb_2O_5 oxide powder determined in the DTA-TG measurements. A possible explanation is the heterogeneous nucleation of the oxide crystallites on the substrate surface as compared to the homogeneous nucleation in the oxide powder, thus occurring at higher temperatures. A behavior similar to that of Nb_2O_5 can be observed in the Ta_2O_5 layers (on Ti30Nb). After annealing at temperatures of 400 to 600 °C the layers are amorphous. A transformation into the crystalline oxide (orthorhombic) is observed after an annealing at 650 °C. As in the case of Nb_2O_5, the amorphous/crystalline transition temperature of the layer (between 600 and 650 °C) is lower than that determined for the oxide powder of Ta_2O_5 by means of DTA-TG (690 °C). In addition, with both oxides (Nb_2O_5 and Ta_2O_5), TiO_2 (rutile) appears after an annealing at 700 °C because of the reaction of the substrate with oxygen diffusing through the layer.

After an annealing at 400 °C, the TiO_2 layer consists of crystalline TiO_2 in the anatase structure (tetragonal), which begins to transform to the tetragonal rutile structure at a temperature of 550 °C. Because of the transformation, the content of anatase decreases step by step with increasing temperature while the content of rutile increases. At 400 °C the ZrO_2 layers are crystalline with a cubic lattice. At 600 °C, besides the cubic ZrO_2, monoclinic ZrO_2 can be detected in the layer. In addition to

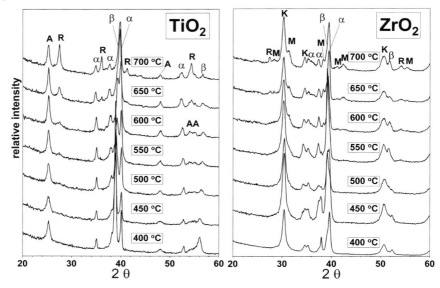

Figure 4.3 Glancing angle XRD pattern of TiO$_2$ (on Ti and ZrO$_2$ sol–gel oxide layers heat treated at different temperatures for 1 h (A: anatase; R: rutile; K: cubic ZrO$_2$; M: monoclinic ZrO$_2$; α: α-phase (titanium); β: β-phase (titanium)).

these phases, at 650 °C rutile is found, which is a result of the thermal oxidation of the metal substrate at the high annealing temperature. As an example for the XRD measurements, the patterns of the TiO$_2$ and the ZrO$_2$ layers on Ti15Mo5Zr3Al and Ti13Nb13Zr, respectively, are shown in Figure 4.3.

Morphology of the Oxide Layers
Important parameters of the oxide morphology are the crystallographic orientation of the oxide grains (e.g. texture) and the shape of the crystallites (e.g. spherical or column). The structure and the morphology of the oxide layers were determined by means of TEM. Figure 4.4 shows a TEM micrograph of a Nb$_2$O$_5$ sol–gel layer with a layer thickness of 158 nm annealed at 600 °C for 1 h and deposited on a foil of cp-Ti, grade 2. On the left-hand side of the micrograph, two metallic grains of the substrate can be observed. Because of their different grain orientations they show different gray scales (Bragg). The oxide layer is located between the metallic substrate and a platinum layer which was used for the TEM preparation procedure. The oxide crystallites, having a size smaller than 50 nm, are spherical in shape and have many different gray scales. This indicates that the layer is not textured and has a statistical distribution of the grain orientations and shows no column growth. The thickness determined from the TEM micrograph amounts to about 160 nm and shows good agreement with the ellipsometric data (158 nm).

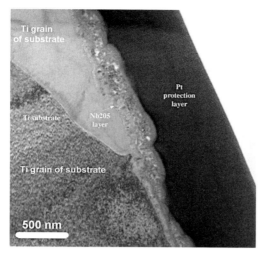

Figure 4.4 TEM micrograph of Nb_2O_5 sol–gel oxide layer on cp-Ti.

Oxide Crystallite Size

The determination of the crystallite size was carried out by means of peak broadening in the glancing-angle XRD pattern, a method developed by Scherrer [1]. The oxide layers were prepared on Ti6Al4V with thicknesses between 121 and 131 nm. The heat treatment was carried out at temperatures above the crystallization temperature of the different oxides for 1 h. Thus, the crystallite sizes for Nb_2O_5 and Ta_2O_5 were determined for temperatures above 550 and 650 °C, respectively. As compared to Nb_2O_5 and Ta_2O_5, the oxides TiO_2 and ZrO_2 are already crystalline at 400 °C and show additionally a phase transition from anatase to rutile and from cubic to monoclinic ZrO_2 respectively at higher temperatures. Up to 550 °C, the crystallite size of these oxides amounts to 10 to 20 nm and increases only a little after an annealing at 400 to 550 °C. With increasing annealing temperature up to 650 °C an increase to values of about 40 nm can be observed. For TiO_2 and ZrO_2, a strong increase of the crystallite size up to 63 and 80 nm, respectively, occurs at 700 °C after the end of the continuous anatase/rutile and cubic/monoclinic (ZrO_2) phase transitions. As an example for the dependence of the crystallite size on the annealing temperature, Figure 4.5 shows this for ZrO_2 sol–gel oxide layers on Ti6Al4V.

Depending on the heat treatment, the size of the oxide crystallites can be varied in a range from 10 to 80 nm. The crystallites grow with increasing annealing temperature according to Ostwald ripening.

Topography

The parameters for the designated adjustment of the topography and the roughness of the oxide surface were determined by means of AFM. The different oxides were deposited on polished samples of Ti6Al7Nb having a surface roughness of $R_a = 4$ nm. The oxide layers were heat treated in a temperature range from 450 to 700 °C for 1 h

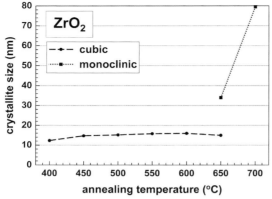

Figure 4.5 Size of the oxide crystallites in the ZrO$_2$ sol–gel films.

and their thicknesses amount to 121 to 131 nm. To determine the influence of the annealing time on the roughness of the oxide surface, TiO$_2$ oxide layers were heat treated at 600 °C for 15 min up to 480 min.

From 400 to 600 °C the roughness R_a increases with increasing temperature from 2 to 20 nm. Above 600 °C the slope in the roughness curve increases and the effect is stronger in the case of TiO$_2$ and ZrO$_2$ than with layers of Nb$_2$O$_5$ and Ta$_2$O$_5$. While TiO$_2$ and ZrO$_2$ coatings reach R_a values of 100 nm and more, the surfaces of Nb$_2$O$_5$ and Ta$_2$O$_5$ show a maximum roughness of about 50 nm. The influence of the annealing time on the roughness is shown in Figure 4.6 for a TiO$_2$ coating.

The measured R_a value increases from 8 nm after an annealing time of 15 min up to 65 nm after 480 min and shows a logarithmic dependence of the roughness on the annealing time.

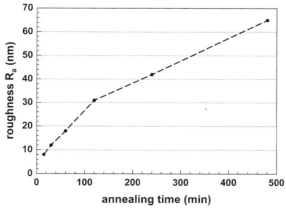

Figure 4.6 Roughness of TiO$_2$ sol–gel surfaces depending on the duration of the heat treatment at 600 °C.

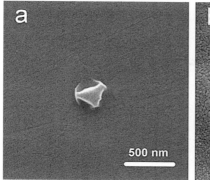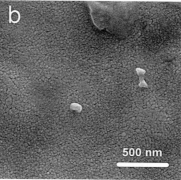

Figure 4.7 HRSEM micrographs of TiO$_2$ sol–gel layers annealed for 1 h at (a) 400 °C, (b) 700 °C.

To visualize the nanosized topography of the sol–gel layers, HRSEM was used. Figure 4.7 shows the surfaces of TiO$_2$ sol–gel layers. The layer in Figure 4.7a was annealed at 400 °C for 1 h having a roughness R_a of 4 nm. The surface is very smooth and no noticeable unevenness can be observed. The oxide crystallites have a size of about 15 nm (determined by XRD) and cannot be visualized by means of HRSEM. Figure 4.7b shows a layer with an R_a value of 104 nm which is the result of a heat treatment at 700 °C for 1 h. In contrast to Figure 4.7a this surface shows a polycrystalline oxide layer where the adjacent oxide crystallites build up the surface. The size of the crystallites amounts to about 50 to 60 nm and is in good agreement with the results from the XRD measurements (63 nm).

Up to this point, all coatings prepared with almost all the same R_a values had been set up to ensure that only the chemical composition of the surface would be responsible for the results achieved. Since the Nb$_2$O$_5$ sol–gel layers showed the best results in the cell viability and cell function tests (Section 3.2 of Part III) as a functional material for enossal implants they were selected for further investigations on the influence of the surface roughness on cell testing. Therefore layers of an identical chemical composition (Nb$_2$O$_5$) were prepared on cp-Ti and were annealed for 1 h at 400, 550 and 700 °C, respectively, in order to vary the topography of the surfaces. The results of this procedure are summarized in Table 4.3. The corresponding AFM

Table 4.3 Thermal treatment, crystallographic phases, roughness and oxide crystallite size of the Nb$_2$O$_5$ sol–gel layers.

	Nb$_2$O$_5$ (1 h 450 °C)	Nb$_2$O$_5$ (1 h 550 °C)	Nb$_2$O$_5$ (1 h 700 °C)
Crystallographic phases	amorphous Nb$_2$O$_5$	crystalline Nb$_2$O$_5$ (monoclinic)	crystalline Nb$_2$O$_5$ (monoclinic)
Roughness, R_a (nm)	7	13	39
Oxide crystallite size (nm)	–	23	58

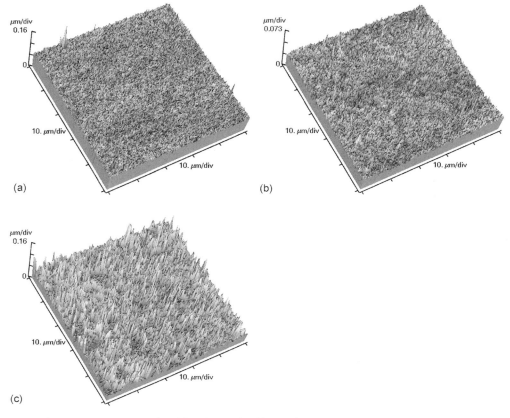

Figure 4.8 AFM micrographs of the Nb_2O_5 sol–gel layers after an annealing of 1 h at (a) 450 °C, $R_a = 7$ nm, (b) 550 °C, $R_a = 13$ nm, (c) 700 °C, $R_a = 39$ nm.

micrographs are shown in Figure 4.8a–c. For a better visualization of the differences in the roughnesses the magnification of the AFM micrographs was kept within the dimension of the cells (20–40 μm). HRSEM micrographs of these layers are shown in Figure 4.9.

Changing Topography by Adding Particles to the Sol

By adding TiO_2 or hydroxyapatite powder to the sol before coating, it is possible to produce particle-filled titania coatings [4]. TiO_2 particles in the coating influence the morphology of the surface by the amount and size of the particles added to the sol (Figures 4.10 and 4.11). Additional to the topography hydroxyapatite particles lead to a calcium ion release. Both modifications influence the cell behavior. For producing particle-filled coatings it is fundamental to avoid agglomeration of the particles by adding an adapted dispersant to the sol.

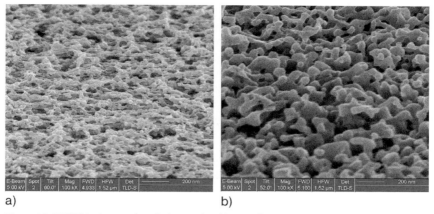

Figure 4.9 HRSEM micrographs of Nb_2O_5 sol–gel layers after an annealing of 1 h at (a) 550 °C, $R_a = 13$ nm, (b) 700 °C, $R_a = 39$ nm.

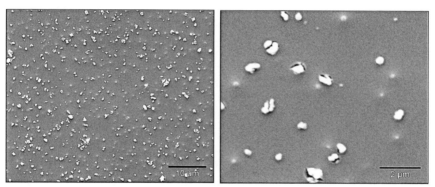

Figure 4.10 SEM images of glass surfaces with TiO_2 coating containing TiO_2 particles (5 wt% TiO_2 powder).

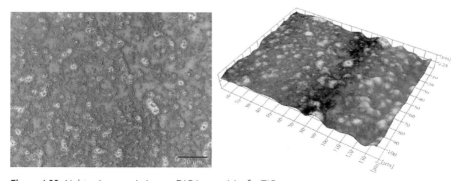

Figure 4.11 Light microscopic image DIC-image (a) of a TiO_2 coating containing 1 wt% TiO_2 particles on a glass disc and the three-dimensional reconstruction of the surface (b).

Wettability

The contact angles of bi-distilled water on different oxide layers were measured by means of the sessile drop method. Because of the strong influence of the topography on the wettability, the investigated layers, which were annealed at 600 °C for 1 h, possess quite the same roughness R_a in the range from 15 to 19 nm. Nb_2O_5 and TiO_2 show the lowest contact angles of 24° and 36°, respectively. The contact angles of Ta_2O_5 and ZrO_2 are 46 nm. For comparison, the contact angles of polished cp-Ti and Ti6Al4V are 41° and 45°, respectively (Figure 4.12).

Metal Ion Release

It is known from literature that dense ceramic titanium oxide coatings can be produced by sol–gel-chemistry via dip-coating or spin-coating techniques [5,6]. Coating was carried out by dipping the substrates in a sol consisting of tetrabutoxy titanate and *n*-butanol with a molar ratio of 1 : 33. The thickness of the layer relates to the pull-out speed (Standard: 1.7 mm s^{-1}). After drying for 1 h in air and calcination at 500 °C, a titanium oxide coating with a thickness of about 140 nm was obtained. Raman spectroscopic measurements indicated pure anatase as the only crystallographic modification (Figure 4.13). Higher calcination temperatures lead to a transformation to the rutile modification. Fourier transform infrared spectroscopy gives evidence that, compared to native passivation layers, the coating possesses a higher number of hydroxy groups at the surface (Figure 4.13). These groups are essential for the surface modification with biomolecules (see Section 4.3).

Patients carrying metallic implants show a significantly higher concentration of metal ions in their blood or surrounding tissue than persons without implants [7,8]. As many metal ions have a toxic or allergic potential, it is essential to reduce the elution of those ions to a minimal level. This leads to another interesting property of titania coatings, the so-called coverfunction. This refers to the fact that under physiological conditions, applying such a ceramic coating as a

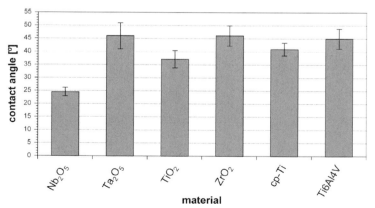

Figure 4.12 Contact angle of bi-distilled water on coated substrates of cp-Ti and on reference materials (cp-Ti, Ti6Al4V).

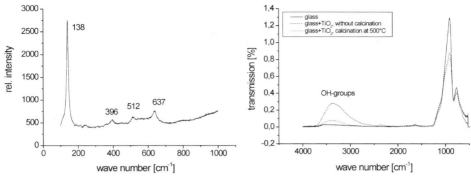

Figure 4.13 (a) Raman spectroscopic measurements of titanium oxide coating (anatase modification); (b) FTIR-ATR spectroscopic measurements of titanium oxide coating. Both specimens were calcined at 500 °C.

diffusion barrier markedly reduces ion release from metallic implant materials into body fluids. To quantify the efficiency of this coverfunction, ion release from Ti6Al4V and Co28Cr6Mo alloys with and without titanium oxide coating was measured as a function of time. The ion release was determined by storing the substrates in a physiological phosphate buffer solution (PBS) for up to 6 months (Figure 4.14) [9]. In the case of the Ti6Al4V alloy, the release of titanium and vanadium ions was significantly reduced by the coating. No significant reduction in the release of aluminum ions could be detected. Compared to titanium and vanadium ions, aluminum ions differ in both ion radius and dielectrical constant. This leads to a higher mobility of the aluminum ions [10]. In the case of Co28Cr6Mo alloy the TiO_2 coating reduces the concentration of all three kinds of ions: Co, Cr and Mo.

Corrosion Properties

To determine the influence of the different sol–gel oxide layers on the corrosion resistance of the Ti30Nb substrates, polished Ti30Nb platelets were coated with different oxides and current density–potential curves were recorded in 0.9% NaCl solution. The thickness, the topography and the heat treatment were the same as in the wettability tests. All coated samples show a broad passive range from 0 to about 1400 mV with a low passive current density and a high polarization resistance. The determined passive current density i_P, breakdown potential U_B and polarization resistance R_P are summarized in Table 4.4.

In particular, the polarization resistance of the different oxides is much higher than that of the Ti30Nb reference material with a native oxide surface layer.

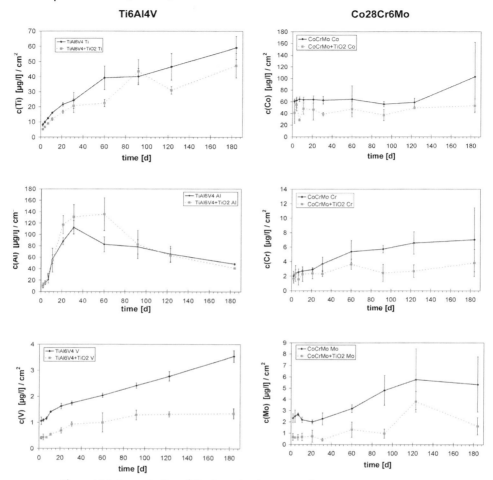

Figure 4.14 Concentration of titanium, aluminum, vanadium, cobalt, chromium and molybdenum ions after different storage times in PBS solution at 37 °C. Ti6Al4V substrate (left), Co28Cr6Mo substrate (right) with (dashed line) and without (solid line) TiO$_2$ coating.

Table 4.4 Corrosion properties of sol–gel-coated Ti30Nb samples in comparison with uncoated Ti30Nb.

	Nb$_2$O$_5$	Ta$_2$O$_5$	TiO$_2$	ZrO$_2$	Ti30Nb Reference
Passive current density, i_P (μA cm^{-2})	0.083	0.062	0.034	0.011	0.6
Breakdown potential, U_B (mV)	1380	1360	1545	1555	1605
Polarization resistance, R_P(MΩ cm^2)	4.82	6.45	11.76	36.36	0.67

4.1.2
Anodic and Thermal Oxide Coatings

Stefan Winter, Dirk Velten, and Frank Aubertin

Oxidation Techniques

Thermal oxidation was performed in a laboratory furnace in air for 5 – 160 min at 600, 650 and 700 °C (Ti15Mo5Zr3Al, Ti13Nb13Zr) and at 600 and 650 °C (Ti30Nb, Ti30Ta).

The electrolyte used for the anodic oxidation was 0.5 M H_2SO_4 at room temperature. To connect the samples as anodes in a galvanostatic circuit, a thin wire consisting of cp-Ti was spot welded to the discs. A commercial platinum electrode was used as counter electrode. Oxygen was permanently bubbled through the electrolyte to adjust a constant concentration during the process. By means of a galvanostat, the current density was kept constant during the oxidation, resulting in a steadily increasing potential. When the required final potential had been reached, the oxidation was terminated by switching off the power supply. Because of the uniform layer growth [11], a current density of 5 mA cm^{-2} was applied and the final potentials at which the oxidation was terminated ranged from 10 to 100 V.

Characterization

To determine the phase compositions of the alloys used, X-ray diffraction (XRD) patterns of polished substrate discs were recorded (Cu K_α, $\Theta - 2\Theta$ scan, $\Theta = $ diffraction angle). The microstructures of the alloys were investigated by means of optical microscopy. The thicknesses of the prepared oxide layers were determined by means of variable angle spectroscopic ellipsometry. This method is suitable because of the transparency of the films in the range of visible light. The Cauchy model was used for fitting the ellipsometric data Ψ and Δ. The Cauchy coefficients for the optical constants of the various oxide layers were verified at three angles (70, 75, 80°). The ellipsometric measurements for determining the thickness of the oxide layers were performed at an angle of 75° at wavelengths from 400 to 750 nm, produced by a xenon lamp.

The elemental compositions of the oxide layers were measured by means of electron spectroscopy for chemical analysis (ESCA) equipped with an Mg K_α and an Al K_α monochromator X-ray source (14 kV, 300 W) and an argon ion sputter gun (4 kV, 2 A m^{-2}). Depth profiles were obtained by alternate cycles of spectra uptake (single peak scans) and subsequent surface removal. The pass energy for single peak scans was 58.7 eV, for survey spectra 187.85 eV. The pressure in the analyzing chamber was maintained at less than 10^{-7} Pa during analysis. The oxidation numbers of the metal cations at the surface were detected by their shape and position in the X-ray photoelectron spectra.

The phases occurring in the oxide layers were detected by means of glancing-angle XRD (Cu K_α) and Fourier transform infrared (FTIR) spectroscopy. The identification of the XRD patterns was achieved by the use of the JCPDS. FTIR analysis was carried out in reflection mode (70°, polarized light) for wavenumbers from 400 to 4000 cm^{-1}. Spectra of polished samples of the various alloys with thin native oxide layers were used as reference.

The corrosion properties of the oxidized samples were determined by measuring current density–potential curves under physiological conditions (0.9% NaCl at 37 °C, pH = 7.4, in oxygen-saturated H_2O). The potential between the working electrode (sample) and the reference electrode (Argenthal Ag/AgCl, +207 mV to the standard hydrogen electrode (SHE)) was raised from −1 to +3 V at a scan rate of 0.1 mV s^{-1}. The instrumental resolution limit of the measured current density is ±5 nA cm^{-2}. The polarization resistance was calculated from the slope of the current density–potential curve in the region of 400–500 mV.

Oxidation Behavior (Ellipsometry)

As determined by means of ellipsometry, the thicknesses of the oxide layers prepared by thermal oxidation depend on the temperature and duration of the heat treatment (Figure 4.15).

Oxide layers with a thickness of 10 to 200 nm can be obtained by thermal oxidation in the temperature range of 600–700 °C. The growth rate increases with increasing

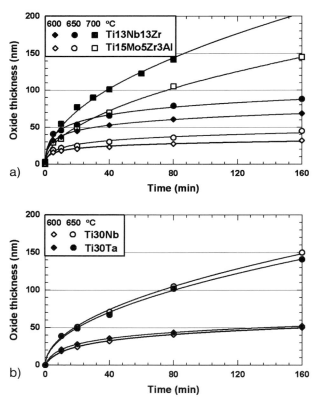

Figure 4.15 Thickness of oxide layers prepared by thermal oxidation as a function of temperature and time: (a) Ti13Nb13Zr and Ti15Mo5Zr3Al, (b) Ti30Nb and Ti30Ta.

temperature and shows a transition from a logarithmic (Eq. 4.1) to a parabolic (Eq. 4.2) behavior:

$$d = k \times ln(t/\text{const} + 1) \tag{4.1}$$

$$d = (2 \times k' \times t)^{1/2} \tag{4.2}$$

where d is the layer thickness, k the logarithmic constant of oxidation, t time and k' the parabolic constant of oxidation. This behavior occurs at temperatures between 650 and 700 °C for Ti13Nb13Zr and Ti15Mo5Zr3Al and between 600 and 650 °C for Ti30Nb and Ti30Ta. As compared with cp-Ti and Ti6Al4V, which already show a parabolic growth rate at 600 °C [11], the oxidation resistance of the β-alloys is increased. The lower oxidation rate is accounted for by the alloying elements (Ta, Nb, Zr, Mo) which decrease the growth of the oxide [12,13]. Ti15Mo5Zr3Al shows the highest oxidation resistance followed by Ti13Nb13Zr and the other near-β-alloys Ti30Nb and Ti30Ta.

Figure 4.16 shows that the measured thicknesses of the anodic oxide films have a linear dependence on the applied final potential, as is reported for cp-Ti and many Ti alloys [11,14].

The growth rates range between 2 and 2.2 nm V^{-1} depending on the alloy used (compared to 2.5 nm V^{-1} (cp-Ti) and 2.1 nm V^{-1} (Ti6Al4V) [11,15]). Using voltages from 5 up to 100 V, the oxide thickness can be adjusted in the range 10–200 nm.

The optical constants of the anodic and thermal oxide layers, measured by means of ellipsometry, are similar for all alloys investigated. Thus the interference colors observed for the oxidized samples can act as an indicator for the layer thickness and provide a possibility of appraising the oxide layer thickness independent of the alloy composition without measurement. The colors and the corresponding thicknesses are listed in Table 4.5.

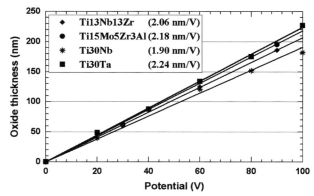

Figure 4.16 Thickness of oxide layers prepared by anodic oxidation as a function of the final potential. The growth rate of the alloys is given in parentheses.

Table 4.5 Colors of oxide layers depending on thickness d.

d (nm)	10–25	25–40	40–50	50–80	80–120	120–150	150–180	180–210
Color	golden	purple	dark blue	light blue	yellow	orange	purple	green

X-Ray Diffraction

The predominant phase in all thermal oxide layers is TiO_2 in rutile structure. Additionally, on the surface of the Ti13Nb13Zr alloy, the orthorhombic $TiZrO_4$ phase ($TiO_2{}^*ZrO_2$ [16]) and the monoclinic $Ti_2Nb_{10}O_{29}$ phase ($2TiO_2{}^*5Nb_2O_5$, [17]), which is also present in the oxide on Ti30Nb, were detected. TiO_2 in the anatase structure was determined in the oxide film on Ti15Mo5Z3Al in addition to the rutile structure, which is the only detectable oxide in the layer on Ti30Ta. Table 4.6 summarizes the detected phases in the thermal oxide layers of the four different alloys.

The thermal oxide layers are mainly rutile because of the excess of titanium in the investigated alloys and the thermodynamic stability of this phase as compared to the other TiO_2 phases, anatase and brookite [11]. The reason for the preferred occurrence of the mixed oxides and the absence of pure oxides of the alloying elements in the thermal oxide layers of Ti13Nb13Zr and Ti30Nb might be explained by the composition of the titanium solid solutions of the bulk metal where niobium or zirconium substitute titanium with a continuous miscibility in the β-phase depending on the temperature. Because of the short diffusion paths in order to build up the mixed oxides during the oxide formation, this process might be favored as compared to a formation of pure oxides. Also the phase diagrams of $TiO_2{}^*Nb_2O_5$ and $TiO_2{}^*ZrO_2$ show the existence of mixed oxides in a wide range of $TiO_2{}^*Nb_2O_5$ and $TiO_2{}^*ZrO_2$ compositions, respectively [18,19]. In the oxide layer on Ti15Mo5Zr3Al, only TiO_2 in anatase and rutile structure is detected, but no molybdenum oxide. In the crystal lattice of TiO_2, the substitution of titanium ions by special foreign ions (e.g. Mo^{6+}) prevents the anatase-to-rutile transformation and stabilizes the anatase structure [20,21]. Thus the molybdenum content in the oxide, determined by means of ESCA, might be in solution in the TiO_2 lattice. Peaks of zirconium- and

Table 4.6 Phases present in the thermal and anodic oxide layers detected by means of XRD.

Alloy	Thermal oxide	Anodic oxide
Ti13Nb13Zr	TiO_2 (rutile) $+ Ti_2Nb_{10}O_{29} + TiZrO_4$	No crystalline oxide phase detected (amorphous)
Ti15Mo5Zr3Al	TiO_2 (rutile) $+ TiO_2$ (anatase)	No crystalline oxide phase detected (amorphous)
Ti30Nb	TiO_2 (rutile) $+ Ti_2Nb_{10}O_{29}$	TiO_2 (anatase)
Ti30Ta	TiO_2 (rutile)	TiO_2 (anatase)

aluminum-containing crystalline phases might be missing because the small intensity of these peaks caused by the low content of these elements in the oxide (<5 at.%)
might be superimposed by the scattering background. Another reason for the
missing peaks of these phases can be that the ions exist in the form of substitutional
ions in the rutile structure. This might also be the reason for no tantalum-containing
phase being detectable in the thermally oxidized Ti30Ta sample.

In the anodic oxide layers, the only detectable crystalline oxide is TiO_2 in anatase
structure on the surface of samples of the alloys Ti30Nb and Ti30Ta. The appearance
of anatase can be attributed to the stabilizing effect of niobium and tantalum cations
on this structure [20,22]. Ti15Mo5Zr3Al and Ti13Nb13Zr show no peaks that can be
assigned to crystalline oxides, revealing that the oxide layers are amorphous or consist
of such small oxide crystallites that the broadening of the XRD peaks leads to the
disappearance of the peaks in the scattering background.

Corrosion Tests

Corrosion resistance, which is an essential prerequisite for the successful use of a
biomaterial in the body, can be evaluated by current density–potential curves which
are shown in the case of the polished samples of the four alloys as compared to
Ti6Al4V in Figure 4.17.

For the polished β- and near-β-Ti alloys with native oxide layers which have a
thickness of 2–5 nm similar to that of cp-Ti [23], the range of passivity with an almost
constant current density between 0.38 and 0.63 $\mu A\,cm^{-2}$ extends from 0 to about
1600 mV. Depending on the alloy the region of transpassivity occurs above 1600 mV
with different breakdown potentials. Potentials of about 300–500 mV, observed with
stainless steels, e.g. 316 L or Co–Cr-based alloys [24], can easily be produced in the
body, e.g. in the case of an inflammation [33]. They lead to an elevated, harmful
current of ions to the host tissue. The corrosion properties of the polished β- and
near-β-alloys are comparable to those of Ti6Al4V [11]. Figure 4.18 shows the current
density–potential curves of the thermally oxidized samples.

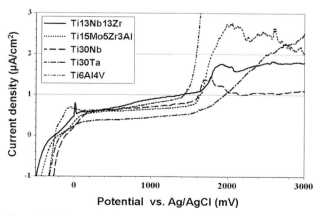

Figure 4.17 Current density–potential curves of the four alloys and Ti6Al4V in the polished state.

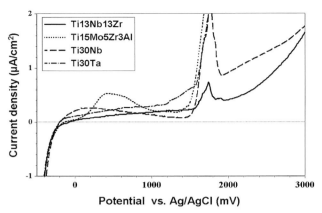

Figure 4.18 Current density—potential curves of the thermally oxidized samples.

The thicker oxide layer (100 nm) enhances the corrosion resistance as compared to the polished samples with a thinner, native oxide layer. The current density can be reduced to about $0.2\,\mu A\,cm^{-2}$ (except for Ti15Mo5Zr3Al: $0.5\,\mu A\,cm^{-2}$), indicating a lower ion dissolution from the thermally oxidized samples as compared to the polished samples, which is also reported for thermal and anodic oxide layers on cp-Ti and Ti6Al4V [11].

After anodic oxidation of the alloy samples resulting in an oxide layer with a thickness of 100 nm the passive current density (Figure 4.19) is decreased to values smaller than $0.01\,\mu A\,cm^{-2}$ which are in the region of the limit of error, revealing excellent corrosion properties and a very small current of ions under physiological conditions.

The measured passive current densities at a potential of 400 mV of the polished and thermally and anodically oxidized samples are listed in Table 4.7.

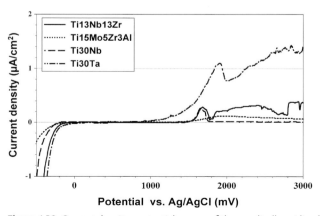

Figure 4.19 Current density—potential curves of the anodically oxidized samples.

Table 4.7 Passive current densities ($\mu A\,cm^{-2}$) at 400 mV/polarization resistance ($k\Omega\,cm^2$) of different oxide layers.

Alloy	Polished (native oxide layer)	Thermal oxidation (thickness 100 nm)	Anodic oxidation (thickness 100 nm)
Ti13Nb13Zr	0.61/656	0.11/3636	0.01/40 000
Ti15Mo5Zr3Al	0.59/677	0.50/800	0.01/40 000
Ti30Nb	0.60/667	0.23/1739	0.01/40 000
Ti30Ta	0.38/1052	0.20/2000	0.01/40 000

Another indicator for a high corrosion resistance is a high polarization resistance which is shown for the different oxide layers in Table 4.4. As compared to the stainless steel 316L and the Co–Cr-based alloy vitallium, which show values of 4.4 and 3.3 $k\Omega\,cm^2$, respectively [24], the polarization resistance of the polished and the oxidized titanium alloys is higher by at least two orders of magnitude.

By thickening the oxide layer, the corrosion resistance of the titanium alloys can be improved as compared to polished samples with their native surface oxide. The passive current density after thermal oxidation is about three times lower than that of the polished samples. By means of anodic oxidation a reduction of the passive current density is obtained down to 0.01 $\mu A\,cm^{-2}$ and the polarization resistance could be raised up to 40 000 $k\Omega\,m$, indicating an increased corrosion resistance.

4.2
Biomimetic Apatite Formation

Frank A. Müller and Lenka Müller

Since biomaterials interact with the body through their surfaces after implantation, the characteristics of the outer surface of the materials are critical properties in controlling the biological response between the biological and the artificial material [26]. Load-bearing implants in direct contact with bone should bond directly to the osseous environment without the formation of an interface layer consisting of fibrous tissue which is a typical feature of bioinert (e.g. titanium and its alloys) materials. Several surface modifications of bioinert materials including physico-chemical [27], morphologic [28] and biochemical ones [29] were investigated to overcome the problems of integration by creating osteoconductive as well as osteoinductive surfaces. The bones of mammals are natural composite materials, where one of the components is an inorganic solid, carbonate hydroxyapatite (CHA). It amounts to 65% of the total bone mass, with the remaining fraction formed by organic matter (mostly collagen) and water. The biological hydroxyapatite contains carbonate ions as substituting groups in both the phosphate (B-type CHA) and hydroxyl (A-type CHA) sites of the HA structure. The carbonate content in the bone mineral (about 3–8 wt%) and the relative amount of substitution in the hydroxyl and phosphate sites depend mainly on the age. The biological inorganic phase was found to be preferentially B-type carbonate substituted. The crystals are nanometer sized, with an average length of 50 nm, 25 nm in width and thicknesses of just 2–5 nm, embedded in the organic matrix. Their small size, low crystallinity and their non-stoichiometric composition are very important factors related to the solubility of biological apatites when compared with mineral apatites [30,31]. Calcium phosphate coatings on bioinert materials are of particular interest, since the bone is composed of carbonated hydroxyapatite. Electrolyte solutions, referred to as simulated body fluid (SBF), reproduce the inorganic part of human blood plasma. Therefore, it can be assumed that the structure of coatings precipitated from SBF will be close to that of biological apatites present in human bone. The interest in applying SBF for preparation of biomimetic apatite coatings has much increased within the last 10 years [32–34]. SBF solutions mentioned in the literature differ mostly in the concentrations of HCO_3^- and Cl^- compared to human blood plasma. In the present chapter, the influence of the HCO_3^- content in SBF on the apatite structure and composition is analyzed using Fourier transform infrared (FTIR) and Raman spectroscopy, X-ray diffraction (XRD) and scanning electron microscopy (SEM). The crystal orientation in the coatings was observed using transmission electron microscopy and selected area electron diffraction pattern analysis (TEM/SAED).

Coating Procedure

Biomimetic apatite coatings were deposited on the surface of chemically pre-treated titanium samples [35] by soaking in SBF solutions under static conditions at 37 °C ± 0.4 °C for 14 days. The pre-treatment of titanium is described in detail in Section 3.2.1.

Five SBF solutions labeled as SBF5, SBF10, SBF15, SBF20, SBF27 with a HCO_3^- content ranging from 5 to 27 mmol L^{-1} and a constant $(Cl^- + HCO_3^-)$ content of 136 mmol L^{-1} were prepared by pipetting calculated amounts of concentrated stock solutions of KCl, NaCl, $NaHCO_3$, $MgSO_4 \cdot 7H_2O$, $CaCl_2$, Tris (tris-hydroxymethyl aminomethane), NaN_3 and KH_2PO_4 into double-distilled water [36]. The theoretical total concentration of ions in each SBF solutions is shown in Table 4.8 and compared with the composition of the inorganic part of human blood plasma. The pH of SBF

Table 4.8 The concentration (mmol L^{-1}) of ions in SBF solutions compared with the composition of the inorganic part of human blood plasma.

	Na$^+$	K$^+$	Ca^{2+}	Mg^{2+}	Cl$^-$	HCO$_3^-$	SO$_4^{2-}$	HPO$_4^{2-}$
SBF5	142.0	5.0	2.5	1.0	131.0	5.0	1.0	1.0
SBF10	142.0	5.0	2.5	1.0	126.0	10.0	1.0	1.0
SBF15	142.0	5.0	2.5	1.0	121.0	15.0	1.0	1.0
SBF20	142.0	5.0	2.5	1.0	116.0	20.0	1.0	1.0
SBF27	142.0	5.0	2.5	1.0	109.0	27.0	1.0	1.0
Human plasma	142.0	3.6–5.5	2.12–2.6	1.0	95.0–107.0	27.0	1.0	0.65–1.45

was adjusted between 7.3 and 7.4 at 37 °C. The samples were freely suspended in a volume of model solution roughly corresponding to a sample surface to volume ratio $S/V = 0.05$ cm^{-1}. After exposure, the samples were washed in double-distilled water.

Theoretical Analysis of the Coating Solutions

It has been described that solutions mimicking human blood plasma are supersaturated with respect to stoichiometric hydroxyapatite [37]. The degree of supersaturation S with respect to different calcium phosphates was calculated using

$$S = (IP/K_{sp})^{1/n} \tag{4.3}$$

where IP is the ionic activity of the compound, K_{sp} the solubility product and n the number of ions in a formula unit. The solubility products of different calcium phosphates at 37 °C are mentioned in Table 4.9 [38–40].

The IP values for compounds listed in Table 4.9 are given by the following equations:

$$\text{for HA:} \quad IP = [Ca^{2+}]^{10}[PO_4^{3-}]^6[OH^-]^2 \gamma_1^2 \gamma_2^{10} \gamma_3^6 \tag{4.4}$$

$$\text{for SCHA (A-type):} \quad IP = [Ca^{2+}]^{10}[PO_4^{3-}]^6[CO_3^{2-}]^{0.5}[OH^-] \gamma_1 \gamma_2^{10.5} \gamma_3^6 \tag{4.5}$$

$$\text{for SCHA (B-type):} \quad IP = [Ca^{2+}]^9[HPO_4^{2-}]^{0.5}[CO_3^{2-}]^{0.5}[PO_4^{3-}]^5[OH^-] \gamma_1 \gamma_2^{10} \gamma_3^5 \tag{4.6}$$

The activity coefficient γ_z of a z-valent ion (e.g. $z = 1$ for OH$^-$, $z = 2$ for Ca^{2+}, etc.) was calculated using the modified Debye–Huckel equation:

$$-\log \gamma z = Az^2[I^{1/2}/(1+I^{1/2})-0.3I] \tag{4.7}$$

proposed by Davies [41]. The constant A is equal to 0.515. The total ionic strength of solution I is given by

$$I = 0.5\Sigma z_i^2 c_i \tag{4.8}$$

where z_i is the charge of the i ion and c_i is its concentration in mol L^{-1} listed in Table 4.8. It was proven that (Eq. 4.7) is valid for ionic strengths up to 0.2 mol L^{-1} [42].

Table 4.9 Solubility products of selected calcium phosphates.

Formula	Name		$-\log K_{SP}$
$Ca_{10}(PO_4)_6(OH)_2$	Hydroxyapatite	HA	117.2
$Ca_{10}(PO_4)_6(CO_3)_{0.5}OH$	Slightly carbonated apatite, A-type	SCHA, A-type	115.6
$Ca_9(HPO_4)_{0.5}(CO_3)_{0.5}(PO_4)_5OH$	Slightly carbonated apatite, B-type	SCHA, B-type	111.54

The concentrations of CO_3^{2-}, HPO_4^{2-} and PO_4^{3-} were calculated using equilibrium constants for the following reactions, respectively [43]:

$$H_2PO_4^- \leftrightarrow HPO_4^{2-} + H^+ \quad pK_3 = 7.18 \tag{4.9}$$

$$HPO_4^{2-} \leftrightarrow PO_4^{3-} + H^+ \quad pK_4 = 12.19 \tag{4.10}$$

$$HCO^{3-} + OH^- \leftrightarrow CO_3^{2-} + H_2O \quad pK_1 = 10.25 \tag{4.11}$$

The values for the degrees of supersaturation in SBF solutions and human blood plasma assuming a physiological pH of 7.4 with respect to the compounds listed in Table 4.9 are depicted in Figure 4.20. The solutions are supersaturated with respect to all phases discussed. While the changes in the degree of supersaturation with respect to HA are negligible, the degree of supersaturation with respect to A- and B-type SCHA varied depending on the content of carbonate ions in the SBF solutions. All SBF solutions are at most saturated with respect to slightly carbonated apatite $Ca_9(HPO_4)_{0.5}(CO_3)_{0.5}(PO_4)_5OH$, where the carbonate ion occupies a phosphate group position in the crystal structure (B-type) [44]. The calculations described above generally indicate that the formation of carbonate-containing apatite from SBF is more likely than the formation of stoichiometric hydroxyapatite.

Characterization of the Biomimetic Coatings
The microstructure and elemental composition of biomimetic coatings precipitated on the sample surface during 2 weeks' soaking in SBF were analyzed by SEM accompanied by energy dispersive X-ray analysis (EDX) on gold sputtered samples.

A representative micrograph is shown in Figure 4.21a. The layers are composed of nanosized plates forming spherical aggregates with a diameter up to 20 µm characteristic for coatings precipitated from solutions mimicking blood plasma [43,45]. Calcium and phosphorus were mainly determined by EDX (Figure 4.21b). From the

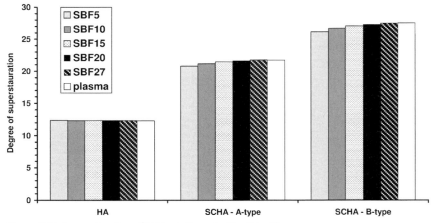

Figure 4.20 Supersaturation of SBF solutions at pH = 7.4 with respect to different calcium phosphates.

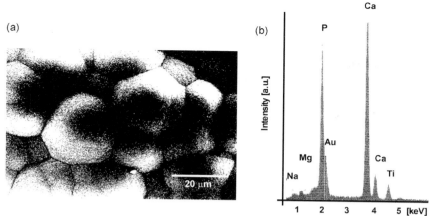

Figure 4.21 Representative (a) SEM and (b) EDX analysis of biomimetic coating precipitated on chemically pre-treated titanium after exposure to SBF10 for 14 days.

traces of Mg^{2+} and Na^+ ions in the spectra it can be supposed that they substitute Ca^{2+} ions. This is in agreement with the composition of biological apatites in human bones which are non-stoichiometric with small fractions of Mg^{2+}, Na^+, CO_3^{2-}, HPO_4^{2-}, F^- or Cl^- [46].

The phase composition of the coatings on chemically pre-treated titanium after exposure to SBF at a physiological temperature of 37 °C was identified as hydroxyapatite of low crystallinity by XRD (Figure 4.22a). The measurement was performed using CuK_α radiation at a scan rate of $0.75° \text{ min}^{-1}$ over a 2θ range of 20–55°. A preferred c-axis orientation of the deposited crystals perpendicular to the titanium surface can be supposed due to the high relative peak intensities of the (002) and (004) diffraction lines at $2\theta = 26°$ and 53°, respectively, compared to the 100% intensity peak of the (211) plane at $2\theta = 32°$. The broad peak at 32° consists of the reflections from (211) at 31.87°, (112) at 32.18° and (300) at 32.87°. Ti and TiH_2 (resulting from the acid etching in HCl) peaks were observed in all XRD patterns. Since the penetration depth of CuK_α radiation into the sample is higher than the thickness (approx. 10 μm) of the calcium phosphate layers, the Ti and TiH_2 diffractions can be detected in the spectra as well. The crystallite size in the c-direction of CHA was calculated from the (002) peak broadening at 26° in the XRD patterns using the Scherrer equation [47]. It decreased from 26 nm in SBF5 to 19 nm in SBF27 (Figure 4.22b).

The cross-sectional TEM view in Figure 4.23a shows the transformations during all treatment steps. The rough surface of titanium hydride with acute tips and valleys was formed by acid etching (see Section 3.2.1). The NaOH treatment accounts for the efflorescence on the surface. Sodium in the hydrated sodium titanate gel is replaced by calcium during soaking in SBF. A biomimetic apatite coating grew on the gel layer by attachment of Ca^{2+} and PO_4^{3-} ions from SBF. Crystal sizes were directly measured from TEM images. Crystal dimensions from $5 \times 5 \times 20$ to $5 \times 25 \times 200$ nm are similar to those found in mammalian bone ($5 \times 25 \times 100$ nm) [48].

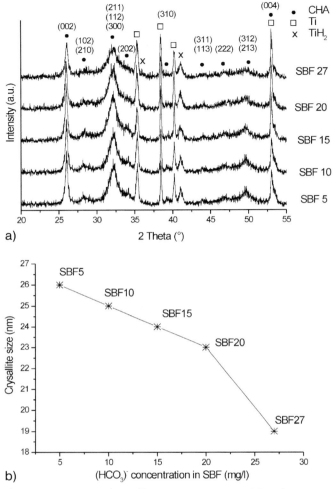

Figure 4.22 (a) XRD analyses of biomimetic coatings precipitated on chemically treated titanium after exposure to SBF5, 10, 15, 20 and 27 for 14 days; (b) dependence of Ca–P crystallite size in the precipitated layers on the HCO_3^- concentration in the SBF solutions.

The corresponding SAED pattern, taken from the area at the top of Figure 4.23a, is shown in Figure 4.23b [49]. It exhibits a large number of rings typical of a nanocrystalline structure. SAED is a technique that allows one to obtain local crystallographic information with high spatial and *d*-spacing accuracy from an area of $1\,\mu m^2$ or even less. In particular, rings which are located close to each other, like 6–7 and 8–9–10, can be resolved. All distances correspond exactly to the hexagonal structure of HA with the parameters $a = 0.944\,nm$ and $c = 0.688\,nm$ according to JCPDS 9–432. Ring 5 corresponds to the basal plane (002).

(a) (b)

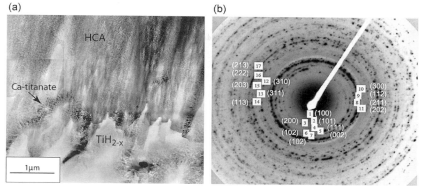

Figure 4.23 Biomimetic growth of carbonated apatite on chemically pre-treated titanium: (a) cross-sectional TEM view from the titanium hydride (bottom) to the HCA layer (top); (b) corresponding SAED pattern taken from the top of the hydroxyapatite layer.

FTIR spectra were measured in transmission using the KBr technique in the range from 4000 to 400 cm^{-1} at a resolution of 4 cm^{-1}. The FTIR spectra of the biomimetic coatings formed after 14 days in SBF5, 10, 15, 20 and 27 are summarized in Figure 4.24. A broad absorption band at 3450 cm^{-1} and the bending mode at 1650 cm^{-1} originating from H_2O were observed in the spectra of precipitated coatings. A broad phosphate band derived from the P–O asymmetric stretching mode (v_3) of the $(PO_4)^{3-}$ group was found in the region from 1200 to 960 cm^{-1} indicating a deviation of the phosphate ions from their ideal tetrahedral structure [50]. The triple (v_4) and double (v_2) degenerated bending modes of O–P–O phosphate bonds were found at

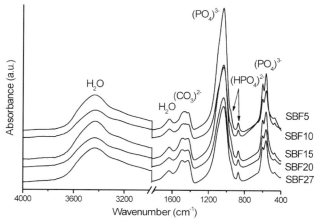

Figure 4.24 FTIR analysis of biomimetic coatings precipitated on chemically pre-treated titanium after exposure to SBF5, 10, 15, 20 and 27 for 14 days.

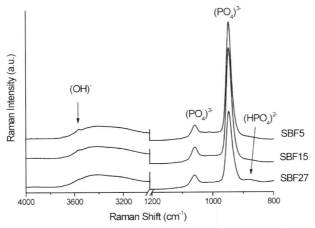

Figure 4.25 Raman spectra of biomimetic coatings precipitated on chemically pre-treated titanium after exposure to SBF5, 15 and 27 for 14 days.

604, 567 and 474 cm^{-1} [51]. The bands detected at 1460, 1420 and 875 cm^{-1} were assigned to the CO_3^{2-} group of B-type carbonated apatite where PO_4^{3-} groups are substituted by CO_3^{2-}. However, the characteristic peaks at 875 and 959 cm^{-1} indicate as well the presence of HPO_4^{2-} in the crystal lattice. Additionally, the absence of the usually sharp vibration band derived from hydroxyl ions at 3650 cm^{-1} can probably indicate the substitution of hydroxyl ions by carbonate ions in the apatite structure (A-type).

The Raman spectra of the coatings were collected in the range from 4000 to 400 cm^{-1} using a dispersive Raman spectrometer equipped with a 780 nm laser. The presence of calcium carbonate in the biomimetic apatite layers can be excluded by the well-discriminated Raman shifts of calcium carbonate and calcium phosphate. None of the characteristic calcium carbonate Raman peaks nor the marker bands of aragonite or vaterite [52] were found in the spectra recorded, even for the calcium phosphate layer precipitated from SBF27 with the highest content of HCO_3^- (Figure 4.25). The band at 3567 cm^{-1} in the SBF5 spectrum is assigned to OH^- group in hydroxyapatite. The phosphate bands were assigned to v_3 (around 1050 cm^{-1}) and v_2 (at 950 cm^{-1}) vibrations [53,54]. The shoulder at 880 cm^{-1} indicates HPO_4^{2-} incorporation into the structure. Similar spectra were recorded for the Ca–P layers precipitated in SBF10, SBF15 and SBF20. In SBF27 the peak at 3567 cm^{-1} disappeared. At the same time the intensity ratio of the bands at 1050 and 950 cm^{-1} decreases with increasing carbonate content in the solution as a consequence of increasing $(CO_3)^{2-}$ substitution for $(PO_4)^{3-}$ (B-type).

Ionic Substitution in Biomimetic Apatites

The composition of SBF is similar to that of the inorganic part of human blood. Therefore, it can be assumed that the structure of the crystalline phase precipitated on the surface of bioactive materials would be close to biological apatite present in human bones. This suggestion was supported by the calculations on the degree of supersaturation with respect to CHA. Bone contains carbonated apatite with CO_3^{2-} substitution of up to 8 wt% [53]. Carbonate ions can be present in the apatite lattice at two different sites replacing OH^- (A-type) and/or $(PO_4)^{3-}$ (B-type). Generally, it is believed that biological apatites are B-type substituted [55]. However, some authors showed that biological apatites are not hydroxylated due to the missing marker band of OH^- in the Raman spectra and thus assume a full A-type substitution [56]. It was also described that the presence of carbonate ions in synthetic apatite influences the decomposition behavior, sinterability, solubility and biological reactivity [55,57].

The FTIR, Raman and X-ray analyses of the precipitated layers indicate that the HCO_3^- content in SBF influences the composition and structure of the resulting calcium phosphates (Figures 4.22, 4.24 and 4.25).

The spectra in Figure 4.24 demonstrated that all layers are composed of B-type carbonated apatite that gives the bands originating from stretching vibrations of CO_3^{2-} at 1420 and 1460 cm^{-1}. The presence of adsorbed and occluded water was demonstrated by the broad peaks at 3400 and 1640 cm^{-1}, respectively. It can be supposed that carbonate substitutes for OH^- (A-type) due to the missing OH^- vibration at 3570 cm^{-1}. However, there is no evidence for CO_3^{2-} substituting for OH^-, as the characteristic absorption at 1545 cm^{-1} associated with A-type substitution was not observed. It is known that calcium-deficient (Ca-def) HA also shows weaker stretching and librational bands compared to stoichiometric apatite [54]. An interesting comparison can be drawn using the Raman spectra in Figure 4.25. Apatite precipitated in SBF5 can be assumed to be hydroxylated due to the presence of OH^- characteristic peak at 3567 cm^{-1}. Since the HCO_3^- concentration in the solutions increases, the band becomes less pronounced and in SBF27 disappears completely probably due to $(CO_3)^{2-}$ substitution for OH^- (A-type).

At increasing carbonate content in the solution, the spectra in Figure 4.24 show a broadening of the v_3 $(PO_4)^{3-}$ band at 1040 cm^{-1} due to higher substitution and thus higher disorder of the phosphate tetrahedron. At the same time, the stretching vibrations of CO_3^{2-} increase in intensity as a consequence of a greater amount of carbonate incorporated in the apatite structure. Furthermore, the carbonate vibrations at 1420 and 1460 cm^{-1} become less resolved and a new vibration appears at 1480 cm^{-1}. According to the literature [61,58] this band is attributed to a carbonate ion in a second unspecified location or to the minor A-type carbonate substitution in B-type CHA [59].

The ratio of the absorption bands at 1420 and 620 cm^{-1} can be used to compare the amount of PO_4^{3-} ions substituted by CO_3^{2-} ions in the precipitated apatites. The $A(CO_3^{2-})/A(PO_4^{3-})$ ratio of the precipitates formed in SBF decreased linearly with decreasing HCO_3^- concentration in the solution. The CO_3^{2-} concentration y in the biomimetic apatite can be calculated according to [60] using

$$y = 10.134x + 0.2134 \qquad (4.12)$$

where x is the ratio $A(CO_3^{2-})/A(PO_4^{3-})$ of the absorption bands. The calculated CO_3^{2-} content in the coatings is given in Figure 4.26. The Ca–P layer formed in SBF27 ($A(CO_3^{2-})/A(PO_4^{3-}) = 0.92$) contained the highest amount of 9 wt% CO_3^{2-}.

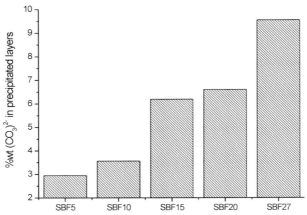

Figure 4.26 Carbonate content in the precipitated CHA calculated form the ratio of the FTIR absorption bands.

The general formula for Ca-def HA can be written as $Ca_{10-x}(HPO_4)_x$ $(PO_4)_{6-x}(OH)_{2-x}$, with $0 \leq x \leq 1$. For stoichiometric hydroxyapatite ($x=0$) the molar Ca/P ratio is 1.67. If the HPO_4^{2-} is incorporated into the structure, the Ca/P ratio decreases to a minimum of 1.5 at $x=1$. CO_3^{2-} can easily substitute for HPO_4^{2-} due to the same charge resulting in a molar Ca/P ratio equal to 1.8 for $x=1$ and full substitution of HPO_4^{2-} by CO_3^{2-}. It is known that Mg^{2+} can substitute up to 10% of Ca [61]. Thus, the formula can be written as: $Ca_{10-x-y}Mg_y(HPO_4)_{x-z}(CO_3)_z(PO_4)_{6-x}$ $(OH)_{2-x}$, where $0 \leq x, y, z \leq 1$. In this case the molar Ca/P ratio ranges between 1.33 (x, $y=1$ and $z=0$) and 1.6 (x, $y=1$ and $z=1$). For the maximum carbonate substitution ($z=1$) the carbonate content is equal to 6.4 wt%. The carbonate content in apatite layers determined using Eq. 4.12 was 2.5, 3.2, 5.7 and 6.2 wt% for SBF5, 10, 15 and 20.

Using FTIR and Raman analyses it was shown that with increasing amount of HCO_3^- in the solution, the substitution of CO_3^{2-} for OH^- occurred and thus the formula can be written as: $Ca_{10-x-y}Mg_y(HPO_4)_{x-z}(CO_3)_z(PO_4)_{6-x}(OH)_{2-x-w}$ $(CO_3)_{w/2}$. The theoretical maximum amount of carbonate substitution in this AB-type CHA for $w=1$ is equal to 10 wt%. In our case, the maximum carbonate content calculated from (Eq. 4.12) amounts to 9.1 wt% in the layer precipitated from SBF27. Hence, it can be assumed that the A-sites are occupied after the B-sites are completely substituted.

According to [32], HCO_3^- inhibits apatite crystal growth, which corresponds to our calculations on Ca–P crystallite size that decreases with increasing HCO_3^- in the solutions. The size of CHA crystallites formed in SBF5, 10, 15 and 20 ranged between 26 and 23 nm. A significantly smaller crystallite size was calculated for apatite precipitated from SBF27 with the highest HCO_3^- content (Figure 4.22b). It is mentioned in the literature [57] that a decrease of the A-type substitution in AB-type CHA caused an increase in the crystallinity of the synthesized powder. The peaks in all XRD patterns (Figure 4.22a) are not sharp and not well resolved. However, the higher relative peak intensity of the (002) and (004) diffraction lines at $2\theta = 26°$ and $53°$ can be attributed to the hexagonal crystal form of hydroxyapatite (HA) with a preferred c-axis orientation of the deposited crystals. Some shifts of peak positions can be observed, reflecting a change in the unit cell dimension due to incorporation of sodium, magnesium and carbonate ions that were detected using EDX (Figure 4.21b). Furthermore, deformation of the PO_4 tetrahedron due to changes in the O–P–O angles was observed, when the quantity of carbonate included in the apatite lattice increased [62]. In our case, it can be supposed that as long as the HCO_3^- concentration in the testing solutions is lower than 20 mmol L^{-1}, only B-type CHA precipitates. At higher HCO_3^- concentration, it can also be assumed that A-type CHA forms considering FTIR, Raman and X-ray measurements.

SBF solutions with different contents of chloride and hydrogen carbonate ions that simulate the inorganic part of human blood plasma were examined. The calculations of degree of supersaturation indicate that the formation of carbonate-containing apatite from SBF is more likely than the formation of stoichiometric hydroxyapatite. The FTIR, Raman and X-ray analyses of the precipitated layers indicate that the HCO_3^- content in SBF influences the composition and structure of calcium phosphates obtained. It can be supposed that as long as the HCO_3^- concentration in the testing solutions is lower than 20 mmol L^{-1}, only B-type CHA precipitates. At higher HCO_3^- concentration, it can also be assumed that A-type CHA forms.

4.3
Surface Modification by Covalent and Noncovalent Attachment of Biomolecules

Chemical modification of an implant surface can be realized, for example, by coating the implant with calcium phosphate using the plasma-spray technique or by wet chemical deposition. Besides such an inorganic surface modification, organic surface modification is also possible. Different strategies for covalent [63–65] or adsorptive [63] surface functionalization with biomolecules were developed. The choice of biomolecules depends on the location and the function of the implant in the body. Both cp-Ti and Ti6AL4V alloys possess a surface oxide layer due to the natural passivation. In addition, hydroxy groups are located at the outermost surface. The thickness of the oxide layer can be increased, e.g. by anodizing [6] or by sol–gel coating of TiO_2. Hydroxy groups are starting points for chemical modification with a heterobifunctional reagent, in which one functional group reacts with the hydroxy group and the second functional group (amino-, carboxy- or carbonate-ester) can react with a spacer. 3-Aminopropyltriethoxysilane (APTES) is one popular reagent for silanization to generate amino groups at the surface. The solvent used and the reaction temperature applied have substantial influence on the resulting concentration of amino groups. When using ethanol/water mixtures, a lower amino group concentration is achieved as compared to pure ethanol or toluene [66]. Figure 4.27 shows different functionalization routes for covalent attachment of bioactive molecules, which were used in our investigations.

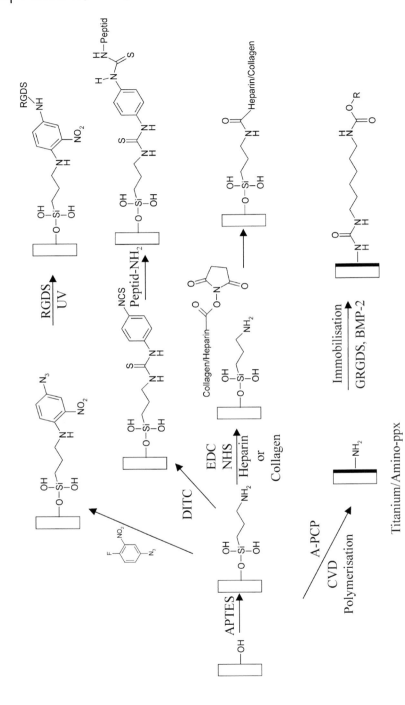

Figure 4.27 Reaction scheme for immobilization of biomolecules. DITC = *p*-phenyleneisothiocyanate, APTES = 3-aminopropyltriethoxysilane, EDC = 1-ethyl-3-dimethylaminopropylcarbodiimid, NHS = *N*-hyroxysuccinimide.

Materials for the implantation into hard tissue (bone) are currently functionalized with proteins that originate from the extracellular matrix (fibronectin, laminin or collagen) [67,68], other proteins found in the body like BMP [69] or oligopeptides containing the RGD motif for cell adhesion [65,70–72]. The aim of this functionalization is the promotion of cell growth. Materials for implantation in blood contact, like stents, can be functionalized with heparin or heparane sulfate to optimize hemocompatibility.

4.3.1
Immobilization of RGD Peptides

Bettina Hoffmann, Volker Faust, Günter Ziegler, Irina Dieser, and Georg Lipps

A nonspecific method to immobilize biomolecules is the coupling via a heterobifunctional reagent with an azido group like 4-azido-1-fluoro-2-nitrobenzene (AFNB). In this investigation, Ti6Al4V discs, coated with TiO$_2$, were cleaned using ultrasound and different solvents (tenside, distilled water, acetone). The surface was functionalized with 2 vol.% 3-aminopropyltriethoxysilane in water/ethanol (1 : 1) at 45 °C for 18 h. The amount of amino groups (4–7 nmol cm^{-2}) found on the surface was afterwards quantified by conductivity titration. In a second step, the amino groups were brought to react with 4-azido-1-fluoro-2-nitrobenzene (50 mg/100 mL) at 37 °C for 24 h in ethanol in the dark. The photochemically active discs were incubated for 15 min with RGDS peptide (1 mg in 3 mL), and then the supernatant was removed. After drying, the discs were exposed to UV light for 15 min (Hg lamp, 150 W). Finally, the discs were washed for 30 min with water, 4 M NaCl solution and distilled water to remove all adsorptively bound RGDS peptide from the surface. To confirm the covalent attachment of the RGDS peptides, time-of-flight secondary ion mass spectrometry (TOF-SIMS) measurements (Tascon, Münster) were carried out. Table 4.10 shows the relative amounts of fragments for selected positively charged secondary ions. The MS spectra of positively charged ions have peaks at mass-to-charge ratios (*m/z*) of 70, 84 and 86, which correspond to ions of the composition $C_4H_8N^+$, $C_5H_{10}N^+$ and $C_5H_{12}N^+$, respectively. These fragments are typical for amino acid sequences located in peptides and proteins [63].

Strategies for Immobilization and Quantification
The adhesion of cells on surfaces is influenced by the properties of the surface and can be improved by biomolecules that promote cell adhesion, e.g. RGD-containing

Table 4.10 Relative peak intensity from TOF-SIMS measurements. Mass fragments of the positively charged secondary ions.

Fragment	*m/z*	Ti6Al4V + TiO$_2$	Ti6Al4V + TiO$_2$ + APTES + AFNB + RGDS
$C_4H_8N^+$	70.069	0.8	7.76
$C_5H_{10}N^+$	84.089	1.31	3.07
$C_5H_{12}N^+$	86.102	0.76	2.37

Table 4.11 Coupling reagents that can react with surface hydroxy groups.

Reagent	Introduced reactive group	Density (nmol cm^{-2})
Hexamethylendiisocyanate	Amino groups[a]	1.6–2.2
3-Aminopropyltriethoxysilane (APTES)	Amino groups	2.6–11
2-Carboxyethylsilantriole	Carboxy groups	5
4-Nitrophenylchlorformiate	Reactive ester	–

[a]After alkaline saponification.

peptides. To immobilize these peptides, reactive groups must be present on the surfaces. Native titania surfaces, which have an oxidized surface layer, and sol–gel-based titanium surfaces have hydroxy groups on their surface. These can be used to introduce coupling reagents that have a higher reactivity than the native hydroxy groups. First, a heterobifunctional coupling reagent reacts with the hydroxy groups present at the surface, and then the second functional group of the coupling reagent (amino, carboxy or a reactive ester group) is used to bind biomolecules either directly or indirectly via a spacer (Table 4.11). In our experience, the silanization reagent APTES is particularly suited to produce amino groups at a high density (up to several nmol cm^{-2}) on the surface of glass or titanium (Figure 4.27).

Initial experiments have revealed that especially the verification of a successful immobilization of biomolecules is technically challenging, since only small amounts of biomolecules are usually immobilized. Therefore, the immobilization is hardly detectable by spectroscopic methods such as X-ray photoelectron and Raman spectroscopy. For that reason we have opted for a different strategy. We do not directly detect the biomolecules on the surface. We have rather chosen an immobilization route which enables us to selectively detach the biomolecule. The biomolecule can then be measured accurately in solution. To this end, we used *p*-phenylenediisothiocyanate (DITC). This reagent reacts at good yields in organic solvents with the primary amines present on the surface and forms a thiourea bond. The second reactive isothiocyanate group is then used to immobilize peptides on the surface via their N-terminal amino group (Figure 4.28). The advantage of this reagent lies in the fact that the immobilized peptide can be selectively detached from the surface by the treatment with trifluoroacetic acid in an Edman peptide sequencing-like degradation. In this reaction scheme, the N-terminal amino acid remains attached to the surface but the rest of the peptide is released. When a fluorescently labeled peptide is used, direct fluorometric quantification of the peptide is possible. For peptides without label, the N-terminal amino group of the peptide can be conveniently used to selectively attach a fluorophore, such as 9-fluorenylmethoxycarbonyl (Fmoc), which then allows for a highly sensitive detection of the peptide by HPLC.

Development of a Method to Immobilize a Fluorescent Model Peptide

In order to establish and optimize the immobilization strategy via *p*-DITC we first used the fluorescently labeled peptide Ala-Ala-Phe-AMC, which carries the fluorophore 7-amido-4-methylcoumarine (AMC). To quantify the peptide we developed a

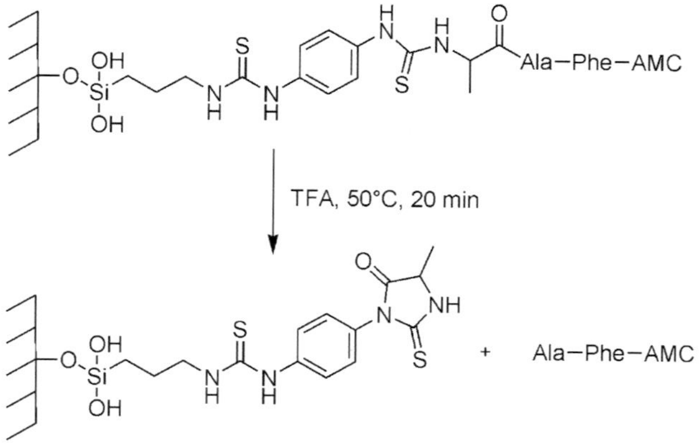

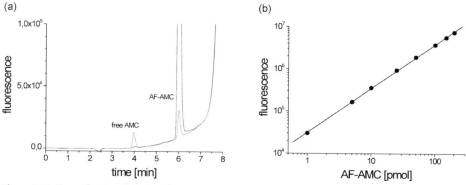

Figure 4.28 Selective detachment of the model peptide Ala-Ala-Phe-aminomethylcoumarine (AAF-AMC) under relatively mild reaction conditions, i.e. 20 min trifluoroacetic acid (TFA). The shortened peptide AF-AMC can be highly sensitively detected by HPLC.

HPLC method, which separates AF-AMC from unspecific signals present in the samples and allows for a precise and reliable quantification of the released peptide from the surface (Figure 4.29). The detection of 5 pmol peptide is easily possible.

On the basis of this powerful detection method, optimal reaction conditions for the immobilization of the peptide on glass and TiO_2-coated surfaces were established. On TiO_2 coatings (approximately $4\,nmol\,cm^{-2}$ amino groups), we were able to immobilize up to $1.5\,nmol\,cm^{-2}$ of model peptide (Figure 4.30). There is a linear relationship between the peptide concentration used for immobilization and the

(a)

(b)

Figure 4.29 Quantification of AF-AMC by HLPC. (a) Chromatogram of 5 and 100 pmol Ala-Phe-AMC respectively on a *reverse phase* C18 column. The fluorescence was excited at 346 nm and detected at 442 nm. (b) Calibration curve over more than two orders of magnitude for the AF-AMC determination. The detection limit of the method is around 1 pmol AF-AMC.

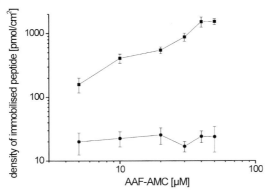

Figure 4.30 Immobilization of the model peptide on TiO$_2$-coated glass surfaces. The model peptide AAF-AMC was immobilized under optimum reaction conditions on TiO$_2$-coated glass surfaces. To quantify the immobilized peptide, TFA was applied, which selectively detaches the peptide. The TFA solution was dried *in vacuo*, the peptide dissolved and the amount of peptide quantified by HPLC. The density of the immobilized peptide was between 150 nmol cm^{-2} and 1500 pmol cm^{-2} (squares). The amount of peptide detected in control reaction (without the crosslinking reagent *p*-DITC) was only about 20 pmol cm^{-2} (dots).

peptide density on the surface. It is therefore possible to adjust the peptide density by choosing the peptide concentration during immobilization. The amount of peptide on the surface in control reactions, i.e. without the crosslinking reagent *p*-DITC, is very low at about 20 pmol cm^{-2}. This amount represents peptide molecules that are not covalently bound but are only adsorbed onto the surface.

Immobilization of the Adhesion Peptide GRGDS

After successful immobilization of the model peptide, the immobilization of adhesion peptides that possess the RGD motif was undertaken. The immobilization chemistry is exactly the same as for the model peptide. However, in contrast to the model peptide, the adhesion peptide is not fluorescently labeled. In order to detect the adhesion peptides at the expected low concentrations, it was therefore required to label the peptide with a fluorophore to allow for a highly sensitive detection. For immobilization, we used an adhesion peptide with the amino acid sequence GRGDS. Through detachment with TFA the peptide is shortened by the N-terminal amino acid to RGDS (see Figure 4.28). The released peptide has an N-terminal amino group, which was used to react with Fmoc. The fluorescently labeled peptide can now be quantified by HPLC (Figure 4.31). This method allows us to detect the adhesion peptide down to about 1 pmol. We used this method to quantify the adhesion peptide immobilized onto glass and TiO$_2$-coated surfaces.

Although our data clearly show that we could immobilize the adhesion peptides on glass and TiO$_2$-coated surfaces, the density of immobilized peptide is disappointingly low. With concentrations of 200–600 μM peptide in the immobilization solution we only achieve a surface density of 3–8 pmol cm^{-2}. In comparison, we achieved a density of approximately 400 pmol cm^{-2} with the model peptide AAF-AMC at a

(a)

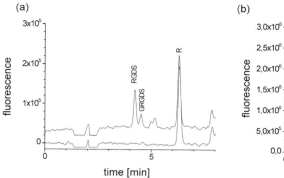

(b)

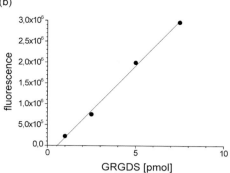

Figure 4.31 Quantification of the adhesion peptide on surfaces. (a) Traces of two samples. The adhesion peptide was immobilized onto TiO₂-coated glass and selectively detached with TFA. The peptide was then spiked with 25 pmol of the internal standard arginine (R), derivatized with FMOC and analyzed on a *reverse phase* C18 column. The detection was performed by fluorescence measurements (excitation: 260 nm, emission 310 nm). The upper trace represents a sample with 4.8 pmol cm^{-2} of RGDS and 0.9 pmol cm^{-2} of GRGDS. The shortened peptide RGDS had been covalently attached to the *p*-DTIC moiety, whereas the unshortened peptide GRGDS is indicative for noncovalently bound adsorbed peptide. The lower trace is a mock reaction without the crosslinking reagent *p*-DITC. The recovery of the internal standard was 97 and 102%, respectively. (b) Calibration curve for RGDS; the detection limit is around 1 pmol.

concentration of only 10 µM. Since both peptides were immobilized with the same chemistry, it appears that the more hydrophobic model peptide AAF-AMC binds more strongly to the surface and reacts more efficiently with the *p*-DITC moiety.

Homogeneity of the Biofunctionalization Using Radioactive Amino Acids for Imaging

For a successful biofunctionalization, the amount and the homogeneity of the immobilized peptide are of central importance. To study how homogenous a biofunctionalization is immobilized onto the *p*-DITC-activated surface we have used ³⁵S-labeled radioactive methionine and ¹⁴C-labeled aspartate, lysine and tyrosine, respectively. One important advantage for using radioactive amino acids is that the same immobilization strategy can be used and that the detection of the immobilization is technically quite simple. After immobilization of the radioactive amino acids the surfaces are washed extensively. The radioactivity in the washing solution was determined by scintillation counting and indicated when all adsorbed noncovalently bound amino acids were washed away. Most importantly, the radioactivity of the immobilized amino acid can be imaged at high sensitivity with an Instant Imager and can also be quantified (Figure 4.32). A homogenous biofunctionalization on glass surfaces was possible when the aminosilanization reaction was performed in the presence of water. On Ti6Al4V, however, we did not achieve a reliable silanization under these conditions. In contrast, under water-free silanization conditions, the

(a)

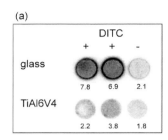

(b)

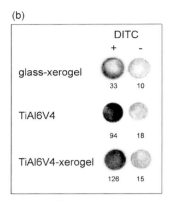

Figure 4.32 Autoradiogram of immobilized ^{35}S-methionin on glass and titania surfaces. The surfaces were treated with APTES in the presence of 5 vol.% H_2O (a) or water-free with toluene (b). Then the surfaces were activated with *p*-DITC (+DITC) or mock-treated (−DITC). Next the surfaces were incubated with 1 nM ^{35}S-methionin in 10 vol.% DMF/90 vol.% pyridine. After washing steps, the amount of immobilized methionine was imaged and the amount of methionine was quantified. The numbers below the images correspond to the density of the amino acid in $fmol\,cm^{-2}$.

silanization was successful and radioactive methionine could be covalently bound to the surface of TiO_2-coated glass, Ti6Al4V and TiO_2-coated Ti6Al4V. However, the homogeneity was less satisfactory than for the silanization of glass in the presence of water in all cases. The inhomogeneity is either caused by the underlying material or by the silanization reaction.

An immobilization strategy for adhesion peptides on glass and titanium surfaces was developed. The method is compatible with the selective detachment of the immobilized peptide and therefore allows for the quantification of the peptide in a low pmol range. In this respect, the methodology developed is superior to other spectroscopic methods to quantify the degree of immobilization. Adaptations of the *p*-DITC-mediated immobilization are valuable for the visualization of the homogeneity of the immobilization by the aid of radioactive biomolecules.

4.3.2
Improving the Biocompatibility of Titanium- and Cobalt-Based Biomaterials by Covalent Immobilization of Crosslinked Collagen

Rainer Müller, Klaus Heckmann, Peter Angele, Richard Kujat, and Michael Nerlich

The aim of the present study was to coat the surfaces of titanium- and cobalt-based biomaterials with layers of fibrillar type-I collagen to improve initial osteoblast adhesion and implant–tissue integration. To enhance stability against enzymatic degradation, the deposited collagen layers were bound covalently to the metal surfaces via silane chemistry and crosslinked by carbodiimide reaction. The coating procedure for biofunctionalization of metallic implant materials realized in our study is illustrated in Figure 4.27. We found that the type of metal used and the oxidation technique applied were key factors for the performance of the collagen coating procedure. For all metallic materials employed, surface biofunctionalization improved *in vitro* adhesion and proliferation of human osteoblast-like cells (MG-63).

Materials and Methods

Disc-shaped and single-side polished specimens of cp-Ti (grade 2), Ti6Al4V, Ti6Al7Nb and Co35Ni20Cr10Mo were used in our investigation. All metal surfaces were chemically etched in 32.5% nitric acid solution for 30 min. Moreover, the surfaces of the titanium materials were electrochemically oxidized in phosphate-buffered solution (pH $= 7.4$) under galvanostatic (260 μA cm^{-2}, 1000 s) or potentiostatic (5 V$_{SCE}$, 2 s) conditions. Elemental analysis of the oxidized surfaces and estimation of the generated surface hydroxyl groups after reaction with gaseous trifluoroacetic anhydride (TFAA) was performed by X-ray photoelectron spectroscopy (XPS). Surface roughness was determined as the arithmetic average peak-to-valley value (R_a) by perthometer measurements [73].

Immobilization of APTES was performed by boiling the oxidized metal specimens in toluene solution containing 15 mg mL^{-1} of APTES for 3 h [73,74]. Surface-bound amino groups were quantified by a colorimetric assay using the dye sulfosuccini-midyl-4-O-(4,4'-dimethoxytrityl)-butyrate (sulfo-SDTB) [75]. Fibrillar collagen was prepared by *in vitro* reconstitution of acid-soluble type I calf skin collagen [76] and dispersed in 2-morpholinoethanesulfonic acid-buffered solution (MES, pH $= 5.5$) at a concentration of 1%. For collagen coating, metal surfaces were immersed in the collagen dispersions and incubated for 4 h after addition of 2.5 mg mL^{-1} 1-ethyl-3-diaminopropylcarbodiimide (EDC) and 0.6 mg mL^{-1} N-hydroxysuccinimide (NHS). The amount of immobilized collagen was determined by a method using sirius red dye in saturated picric acid solution [73,74,77].

Enzymatic stability of the immobilized collagen layers was tested by *in vitro* collagenase digestion. The specimens were incubated at 37 °C with 1 mL of Dulbecco's modified Eagle medium (DMEM) containing 1 mg mL^{-1} collagenase P from *Clostridium histolyticum* for 3, 6, 12 h and 6 days. The enzyme solution was changed daily. Thereafter, the residues of the collagen layers were quantified with picrosirius red [73,74,77].

Coated metal samples were sterilized via β-irradiation (e-beam) with a dose of 25 kGy. Human immortalized osteoblast-like cells MG-63 were routinely cultured in DMEM with high glucose content supplemented with 10% fetal calf serum and penicillin-streptomycin under standard culture conditions (37 °C, 5% CO$_2$). Cells were suspended in serum-containing DMEM and 50 000 cells were plated on each sample surface to determine the adhesion rate which was assessed after 1 day of standard cell culture. For the analysis of cell proliferation, 20 000 cells were incubated and determined after 7 days of culture. The number of attached cells was measured by mitochondrial dehydrogenase activity using 3-(4,5-dimethylthiazol-2-yl)-5-(3-carboxy-methoxyphenyl)-2-(4-sulfophenyl)-2H-tetrazolium inner salt (MTS) and phenazine ethosulfate [73,74].

All results are shown as medians including the 25–75% quantiles.

Table 4.12 Surface elemental composition and roughness of Ti6Al4V after performing three different oxidation procedures.

	XPS of oxidized Ti6Al4V surfaces (at.%)						XPS: TFAA F 1s (counts)	Roughness, R_a (μm)
	C 1s	O 1s	Al 2p	N 1s	Ti 2p	P 2p		
Acid etched	35.1	45.1	2.5	2.1	15.1	–	20 600	0.11 (0.08–0.11)
Galvanostatically oxidized	33.4	46.0	2.0	1.1	9.2	3.4	30 600	0.08 (0.08–0.11)
Potentiostatically oxidized	31.2	46.4	2.4	0.7	10.3	3.2	31 800	0.11 (0.08–0.11)

The relative amount of surface hydroxyl groups can be estimated by the XPS signals for fluorine introduced by TFAA reaction.

Oxidation of Metallic Biomaterials

The influence of the different oxidation procedures on the surface properties of the metals is exemplified for Ti6Al4V (Table 4.12). From XPS measurements no dependency of the applied oxidation technique on the oxygen content of the metal surface could be detected. All surfaces exhibited a high carbon content, whereby adsorption of contaminants from air was made responsible for this effect. However, different contents in surface-bound hydroxyl groups were detected in relation to the oxidation conditions. Electrochemically driven oxidation produced higher amounts of hydroxyl groups on Ti6Al4V than acid etching. The hydroxyl group content was estimated after ester formation with TFAA which leads to an introduction of fluorine to the metal surfaces. Surface roughness at the macroscopic scale displayed no significant difference depending on the oxidation conditions applied.

Silanization of Oxidized Metal Surfaces

Coating of the differently oxidized metals with the silane coupling agent APTES created surfaces of varying amino group content (Figure 4.33). The highest surface concentration of amino groups with 0.75 nmol cm^{-2} was detected on galvanostatically oxidized Ti6Al4V. This biomaterial showed remarkable variations in the amount of immobilized silane molecules depending on the oxidation pretreatment. A medium surface amino group concentration of 0.53 nmol cm^{-2} was obtained when the surfaces were oxidized under potentiostatic conditions. Acid etching led to the immobilization of 0.44 nmol cm^{-2} silane molecules. For cp-Ti, no influence of the applied oxidation technique on the amino group content was found; in all cases an amino group concentration of about 0.57 nmol cm^{-2} was determined. The lowest concentration of immobilized APTES molecules with 0.21 nmol cm^{-2} was detected on the surface of the cobalt alloy.

Immobilization of Fibrillar Collagen

Reconstituted fibrillar collagen was immobilized at the metal surfaces and crosslinked using carbodiimide chemistry. The incubation of oxidized but not

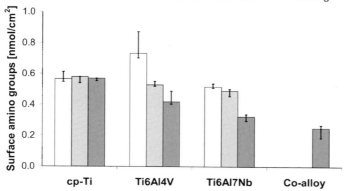

Figure 4.33 Influence of the oxidation pretreatment on the coating of metal surfaces with APTES. Amounts of immobilized silane coupling agent varied with the type of metal used and the oxidation process applied. Surface-bound amino groups were determined with sulfo-SDTB (data shown as medians and 25–75% quantiles, $n = 5$).

APTES-modified metal surfaces resulted in an adsorptive binding of the crosslinked collagen layer. If the metal surfaces were modified with APTES, the covalent attachment of collagen was achieved by binding of carbodiimide-activated carboxylic groups of the protein to the surface-bound amino groups. Adsorptive binding to oxidized cp-Ti resulted in 0.36 (0.26–0.44) $\mu g\,cm^{-2}$ surface-bound collagen. After covalent immobilization to the APTES-modified surfaces higher amounts of collagen have been detected with 1.93 (1.75–2.13) $\mu g\,cm^{-2}$ on Ti6Al4V, 1.56 (1.30–1.73) $\mu g\,cm^{-2}$ on cp-Ti and 1.59 (1.40–1.79) $\mu g\,cm^{-2}$ on cobalt alloy.

Enzymatic Stability of Surface-Bound Collagen

The surface-bound collagen layers were incubated with collagenase solution to study stability against enzymatic degradation (Figure 4.34). The collagen layers, which were physically adsorbed on cp-Ti, did not resist enzymatic degradation, because even after 1.5 h of digestion only 20% of the original amount could be detected on the metal surface. If the collagen was covalently attached to the same metallic substrate, distinctly higher amounts of residual protein were found at each time point of collagenase digestion. The highest resistance of the bound collagen layer was found after covalent immobilization on Ti6Al4V surfaces which exhibited the highest density of silane coupling agents. Collagen layers immobilized on cobalt alloy displayed poor stability against enzymatic degradation. These results indicate an obvious relationship between the surface density of accessible amino groups and the enzymatic stability of the covalently immobilized collagen layer. An increase in the surface density of amino groups led to a higher density of covalent linkages which enhanced the stability of the surface-bound collagen layer.

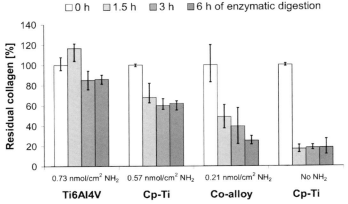

□ 0 h □ 1.5 h ▨ 3 h ▦ 6 h of enzymatic digestion

Figure 4.34 Enzymatic stability of immobilized collagen depending on the degree of surface-supported amino groups. Collagenase-resistant collagen was quantified by sirius red staining after certain incubation periods. Covalent immobilization clearly enhanced resistance against degradation and stability additionally displayed dependencies upon the amount of available surface-bound amino groups (data shown as medians and 25–75% quantiles, $n = 8$).

Cellular Response to Biofunctionalized Metals

Adhesion and proliferation of human osteoblast-like cells was enhanced after coating metallic biomaterials with layers of crosslinked collagen (Figure 4.35). We found that the positive effect on cell behavior did not depend on the type of metal to which the collagen was immobilized. Therefore, the amount of surface-bound collagen and not the stability of the bound collagen layer seems to be the determining factor influencing cell behavior. Each covalently immobilized collagen layer served as a stimulating environment for cell attachment and cell growth.

4.3.3
Modifying Titanium Surfaces with Glycosaminoglycanes for Hemocompatibility

David Tebbe, Uwe Gbureck, and Roger Thull

The aim of the project was an optimization of the hemocompatibility of implant surfaces in contact with blood by means of the modification of the surface topography and composition. The latter was modified chemically by covalent coupling of glycosaminoglycanes (heparin, heparane sulfate) via aminosilane spacers varying the amount of amino groups and the length of the coupling agent. The principle of the surface modification is demonstrated in Figure 4.27 for 3-(trimethoxysilyl)propylamine (APMS) as spacer. Alternative spacer molecules with two or three amino moieties are *N*-[3-(trimethoxysilyl)propyl]ethylenediamine (Diamino-APMS) and N^1-[3-(trimethoxysilyl)propyl]diethylenetriamine (Triamino-APMS).

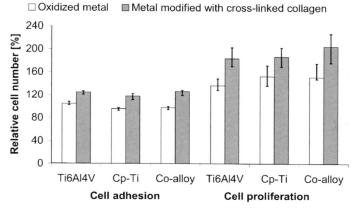

Figure 4.35 Adhesion (1 day) and proliferation (7 days) of osteoblast-like cells on metals, which were collagen-coated by covalent immobilization, compared to oxidized surfaces. Cell behavior was improved by collagen coating on all metal substrates. 100% relative cell number represents the cell culture polystyrene control surface (data shown as medians and 25–75% quantiles, $n = 8$).

Silanisation of Titanium Dioxide and cp-Ti

Modification of TiO_2 was carried out under variation of a published preparation regime [78]. Amounts of 2 g TiO_2 (25.0 mmol) and 2 mL APMS (11.3 mmol) were added under argon atmosphere into 50 mL anhydrous toluene. Then, this suspension was treated with ultrasound (35 kHz) for 2 h. Subsequently, the functionalized TiO_2 was separated with a G4-frit from the solution, washed with toluene and then purified with ethyl acetate in a Soxhlet extractor for 2 h. Finally, the sample was dried in vacuum at room temperature. The reaction was alternatively performed with 2.46 mL Diamino-APMS (11.3 mmol) or 2.91 mL Triamino-APMS (11.3 mmol).

Surfaces of cp-Ti were modified in a similar manner. The substrates were firstly cleaned in 5% EXTRAN solution for 10 min using ultrasound following three washes with deionized water. Oxidation of the metal surface to produce a homogeneous oxide layer was either performed using a solution of conc. H_2SO_4/30% H_2O_2 (1 : 1) for 3.5 h at room temperature or thermally by annealing the substrates at 750 °C for 90 min in a furnace. These samples were then boiled in a solution of 7 mL APMS (39.67 mmol) in 70 mL toluene for 6 h. An elongation of the APMS spacer was performed with adipinic acid. Briefly, 365 mg adipinic acid (2.50 mmol) and 150 mg CMC (0.35 mmol) were dissolved in 50 mL 0.1 M 2-(N-morpholino)ethanesulfonic acid (MES) buffer (5.00 mmol, pH = 4.7 adjusted with 1 N NaOH). The APMS-modified samples were placed in this solution for 17 h at 7 °C following washes with 4 N NaCl and deionized water.

Covalent Attachment of Heparin and Heparan Sulfate

In each case, 500 mg of the functionalized TiO_2 was suspended in 50 mL MES buffer (50 mM, 40% (v/v) ethyl alcohol/water, pH = 5.5) and stirred for 15 min. Then,

95.0 mg N-(3-dimethylaminopropyl)-N'-ethylcarbodiimide hydrochloride (EDC; 0.50 mmol), 12.0 mg N-hydroxysuccinimide (NHS; 0.10 mmol) and 100 mg heparin (~5.60 µmol) were added to the suspensions followed by stirring the reaction mixture at room temperature for 6 h. Finally, the solid was separated by centrifugation and the obtained modified powders were washed three times each with Na_2HPO_4 solution (0.10 M), NaCl-Lösung (2.00 M) and deionized water and then dried in vacuum at 40 °C overnight. The modification of APMS-modified titanium substrates with heparin and heparan sulfate was performed in a similar manner by stirring the metal samples in a solution of 19 mg EDC (0.1 mmol), 2 mg NHS (0.02 mmol) and 20 mg heparin/dermatan sulfate in 50 mL MES buffer for 6 h at room temperature.

Quantification of Surface-Bound Spacer and Heparin

The quantification of the spacer was determined by the amount of terminal primary amino groups and could be measured photometrically with the ninhydrin reaction after Moore and Stein [79]. Basically, this method is based upon the reaction of primary amino groups with ninhydrin to give the dye Ruhemanns Purpur, which can be quantified photometrically at 570 nm. The amount of immobilized heparin was quantified by means of the toluidine blue method, already described in the literature [80]. This method is based on complexation of the dye toluidine blue with heparin in aqueous medium. By the use of an excess of the dye, the residual "free" toluidine blue can be determined photometrically at 631 nm and can be correlated with the amount of heparin.

Hydrolysis Behavior of Immobilized Heparin

Five quantities (each 250 mg) of the TiO_2 powders, modified with APMS, Di- and Triamino-APMS and heparin, respectively, were placed in glass flasks. Then, 5 mL PBS buffer was added and the suspensions were treated with ultrasound for 3 min. The flasks were densely closed with a plastic lid and then given inside an incubator (37 °C) onto a shaker (70 shakes per minute). At scheduled times one quantity (250 mg) of each heparinized sample was taken, the powder was separated from the liquid with a centrifuge and then dried in vacuum. The amount of residual immobilized heparin was quantified photometrically with the toluidine blue method.

Determination of the Biological Potency of Heparin

For the determination of the biological activity of the immobilized heparin, the chromogenic substrate Chromozym TH was used. The principle of the quantification is based upon heparin formation with antithrombin (AT-III), a heparin/AT-III complex in an aqueous medium (Figure 4.36). In the presence of an excess of thrombin, formation of a heparin/AT-III/thrombin complex follows. Excessive thrombin in solution catalyzes the hydrolysis of the chromogenic substrate Chromozym TH into the respective dye, which can be measured photometrically at 405 nm [81]. The higher the biological activity of the heparin remains after immobilization, the more thrombin is complexed and the less dye will be hydrolyzed.

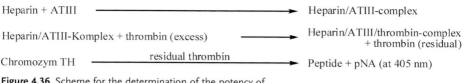

Figure 4.36 Scheme for the determination of the potency of heparin by means of the chromogenic substrate Chromozym TH.

Further Characterization Methods

Raman spectra were recorded with a spectrometer at a stimulating wavelength of $\lambda = 633$ nm (HeNe laser). The laser beam was focused onto the test samples at 50-fold magnification with a PL Fluotar objective. Therefore, the powders were pressed with a glass object holder onto a swirl, diffuse dispersing metal disc. The stimulation was carried out with 50 mW. The data so obtained were analyzed with a software package.

The determination of the zeta potentials was carried out with a Zeta-sizer equipped with a HeNe laser ($\lambda = 633$ nm) and a standard electrophoresis cell at an electrical field force of ± 150 mV.

The adsorption of proteins onto the modified surfaces was measured in real time using quartz crystal microbalance with dissipation monitoring (QCM-D). Modification of the quartz crystals was performed in three steps: a thin TiO_2 coating was applied using physical vapor deposition (PVD) following chemical coupling of APMS and heparin, similar to the methods described for cp-Ti samples.

Results and Discussion

Quantification of the Immobilized Primary Amino Groups from the Spacer Molecules

The covalent bonding of heparin on titanium dioxide was realized by the spacers APMS, Diamino-APMS and Triamino-APMS, which were dissolved in toluene and immobilized on hydroxyl groups of the TiO_2. In a following reaction, activated carboxyl groups with EDC and NHS of heparin formed an amide bonding with terminal amino groups of the spacer (Figure 4.27). Covalent attachment of heparin on Diamino-APMS and Triamino-APMS was also carried out by reaction of primary amino groups ($R–NH_2$) from the spacer with activated carboxyl groups of heparin. Secondary amino groups ($R–NH–R'$) are not able to form amide bondings with carboxyl groups. The amount of covalently attached primary amino groups on the titania surface (powder and discs) was determined photometrically by the ninhydrin reaction and is shown in Table 4.13.

Table 4.13 Quantification of NH_2 groups per area (NH_2 nm^{-2}) for different spacer molecules on cp-Ti disc.

APMS	Diamino-APMS	Triamino-APMS
156 ± 16	162 ± 12	134 ± 22

Table 4.14 Quantification of NH$_2$ groups as amount per powder weight and respectively as amount per area.

TiO$_2$ powder	APMS	Diamino-APMS	Triamino-APMS
nmol mg^{-1}	84.9 ± 2.1	78.7 ± 2.0	71.3 ± 1.3
NH$_2$ nm^{-2}	20.0 ± 0.5	18.7 ± 0.5	17.0 ± 0.3

The amount of immobilized terminal amino groups was the highest for APMS and decreased for Diamino-APMS and Triamino-APMS. This phenomenon can be explained, since the larger molecules like Triamino-APMS are sterically more hindered on the titania surface than the smaller ones. Furthermore, a higher number of secondary amino groups that are protonated in aqueous medium at pH = 7 to R$_2$NH$_2^+$ result in a stronger intermolecular electrostatic repulsion. Both effects lead to less functionalization of the surface with spacer molecules. Comparing the values of Table 4.14 with those of the literature, where a monolayer of NH$_2$ groups is described with 2–3 NH$_2$ nm^{-2}, the aminations accomplished here are in the range of multilayers [82]. Furthermore, the number of amino groups on titanium discs was multiple times higher than those on the powders (Table 4.14). However, calculating the surface area of the discs, ideally plane surfaces were supposed, whereas the high micro-roughness of the real surface was not considered, which makes the actual surface area much higher than expected.

Quantification of Heparin Quantification of immobilized heparin was done by the toluidine blue method and was only applicable for the drug-modified powders (Table 4.15). On the titanium discs, no heparin was detectable by this method, probably because the amount of bound drug was less than the detection limit of this method (10 µg).

On the APMS spacer, the greatest amount of 53.3 ± 3.6 ng cm^{-2} heparin was attached compared to Diamino-APMS and Triamino-APMS (Table 4.15). This is due to the highest density of surface-bound amino groups on TiO$_2$ with APMS that can form bondings with carboxylic groups of the heparin. If the amount of attached heparin is compared with others known from the literature, the values obtained here are much lower (Table 4.16). This is probably due to the fact that polymers can bind heparin over their whole volume which can be considered as a heparinization in the range of very high multilayers. On the one hand, surfaces modified with

Table 4.15 Quantification (ng cm^{-2}) of heparin immobilized on APMS-, Diamino-APMS- and Triamino-APMS-modified titania powders.

APMS	Diamino-APMS	Triamino-APMS
53.3 ± 3.6	41.2 ± 6.9	32.1 ± 5.7

Table 4.16 Amount of immobilized heparin on different polymer substrates.

Substrate	Polyurethane-CO₂H [88]	Polyurethane-NH₂ [88]	Polyurethane-PEO [89]	Polyethylene [90]
Immobilized heparin ($\mu g\,cm^{-2}$)	1.40 ± 0.08	2.00 ± 0.13	0.24 ± 0.04 $0.47 \pm 0.08^{a)}$	4.4 ± 0.1 4.7 ± 0.1

a)Dependent on PEO spacer length.

APMS can bind more heparin than those functionalized with Diamino-APMS and Triamino-APMS. Otherwise, longer spacer molecules like Triamino-APMS have the advantage that the potency of the immobilized heparin is higher due to its higher flexibility.

Reaction Control by Means of the Zeta Potential Determination of the zeta potentials from the unmodified, spacer functionalized or rather heparinized powders showed the successful functionalization of the titanium dioxide (Figure 4.37).

Unmodified TiO_2 has a negative zeta potential of -26.1 ± 10.5 mV due to its deprotonated hydroxyl groups at pH = 7 in water. After modification with the different spacer molecules, the zeta potentials turned positive to $+44.1 \pm 4.4$ mV (APMS), $+40.6 \pm 6.2$ mV (Diamino-APMS) and 45.3 ± 5.5 mV (Triamino-APMS), because of the protonated primary amino groups (RNH_3^+) of the spacer molecules in water. Finally, the covalent attachment of heparin changes the zeta potentials negative to -37.2 ± 0.9 mV (APMS), -39.3 ± 1.0 mV (Diamino-APMS) and -39.0 ± 0.2 mV (Triamino-APMS). The strong decrease of the zeta potentials is caused by negative sulfate and carboxyl groups of the drug in the aqueous medium at pH = 7.

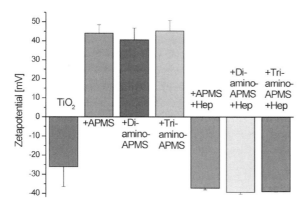

Figure 4.37 Zeta potentials after functionalization of TiO_2 powder with APMS, Diamino-APMS and Triamino-APMS and respectively after the reaction with heparin. Each data point is the mean of three samples; error bars are standard deviations.

Raman Spectroscopy The successful functionalization of titania powder with APMS and heparin could also be followed using Raman spectroscopy (Figures 4.38 and 4.39) by the occurrence of the peaks caused by the C–H vibration at 2800–3000 cm^{-1} and the symmetrical S = O vibration at 1040 cm^{-1}.

The attachment of heparin on titanium dioxide by means of the spacer molecules Diamino-APMS and Triamino-APMS can also be followed using Raman spectroscopy, due to the appearance of the peak at 1040 cm^{-1} from the heparin.

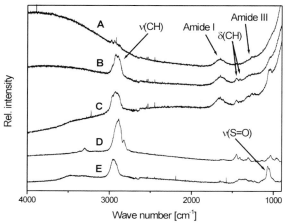

Figure 4.38 Raman spectra of anatase (A) after silanization with APMS (B) and with heparin (C) compared with APMS (D) and heparin (E).

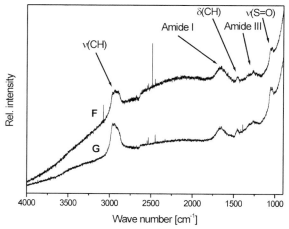

Figure 4.39 Raman spectra of TiO$_2$ after modification with Diamino-APMS and heparin (F) and after the reaction with Triamino-APMS and heparin (G).

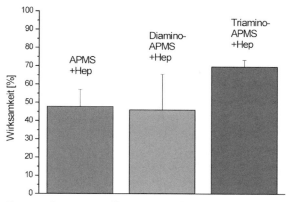

Figure 4.40 Potencies of heparin immobilized by the spacers APMS, Diamino-APMS and Triamino-APMS.

Biological Activity and Hydrolysis Behavior of Immobilized Heparin The potency of the immobilized heparin was measured photometrically with the chromogenic substrate Chromozym TH at $\lambda = 405$ nm. Comparing the potencies of the covalently attached heparin on Diamino-APMS and Triamino-APMS with APMS, it becomes obvious that longer spacer molecules tend to result in higher biological activity of heparin (Figure 4.40). Actually, the value for Diamino-APMS should be higher than that for APMS, because the former spacer is longer. This results in a higher flexibility of the immobilized heparin which causes a higher activity of the drug. However, it has to be noted that each functionalized powder was measured only twice, which results in relatively large deviations.

The influence of the secondary amino groups on the immobilization of heparin was investigated by measuring the hydrolysis rate of the drug under *in vitro* simulated physiological conditions (37 °C, PBS, 70 shakes per minute). In an aqueous medium, these secondary amino groups are protonated to $R_2NH_2^+$, which enables them to interact with negative sulfate, carboxylate and hydroxyl groups of the heparin via electrostatic interactions. This ionic bonding should be weaker than the covalent bonding and should have only an additional effect. Figure 4.41 shows that spacers like Triamino-APMS with many secondary amino groups has the highest hydrolysis rate of the drug during the first 100 h, compared to Diamino-APMS and APMS. This effect is because of heparin molecules that are only bound electrostatically, which makes the hydrolysis easier and faster. After this period, the hydrolysis profile changes and the drug hydrolyzed more slowly from Triamino-APMS than from the others. The reason probably is that the remaining heparin molecules are both covalently and ionically bound on the Triamino-APMS compared to pure APMS, which can only bind the drug covalently.

Protein Adsorption of Heparinized Surfaces The primary reaction after incorporating an implant into the body is the formation of a biofilm composed of adsorbed matrix proteins. The quantitative composition of this protein film and conformational

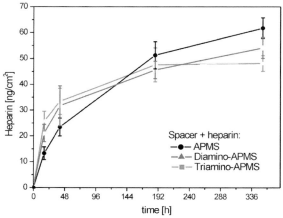

Figure 4.41 Hydrolysis of heparin from the spacer molecules APMS, Diamino-APMS and Triamino-APMS in aqueous medium.

changes during adsorption are thought to be the key factor controlling the attachment of cells and the biological response to the implant material [86]. In the case of materials in contact with blood, it is also known that the adsorption of various proteins combined with structural changes of the protein layer, e.g. fibrinogen can initiate blood clotting. The adsorption of fibrinogen to heparinized TiO_2 surfaces was measured in this study using the QCM-D technique. This method offers two advantages: firstly, the adsorption can be followed in real time due to the mass increase of the quartz crystal; secondly, it is possible to analyze conformational changes due to the viscoelastic properties of the adsorbed protein layer. Therefore, the gold surfaces of the quartz crystals were modified in a first step with a thin (approx. 100 nm) TiO_2 layer using the PVD technique which was then coated with the coupling agent APMS and heparin similar to the methods described for titanium substrates.

Figure 4.42a shows changes of the frequency Δf and the dissipation ΔD for the various surfaces during measurement. The addition of fibrinogen dissolved in PBS buffer occurred after 5 min and resulted in a decrease of Δf and an increase of ΔD. Since Δf correlates with a mass increase of the crystal due the adsorbed protein layer and ΔD depends on the viscoelastic properties of the layer, the results demonstrate that the protein adsorption on gold and TiO_2 behaved similarly. In contrast, the heparinized surface showed a reduced adsorbed mass of fibrinogen as well as a higher elasticity of the protein layer. Figure 4.42b shows a correlation between ΔD and Δf which is helpful for analyzing the viscoelastic properties of the protein films since Δf and ΔD do not have the same time dependency such that time as a parameter can be eliminated using this approach. In various experiments it was shown that a low $\Delta D/\Delta f$ is an indication for rigid adsorbed films [87]. The results show that for both gold and TiO_2 surfaces the $\Delta D/\Delta f$ ratio is similar and constant over a broad frequency range with only a small decrease at higher frequencies.

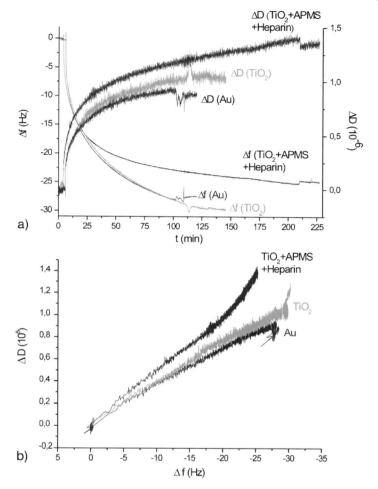

Figure 4.42 (a) Δf (decreasing curves) and ΔD (increasing curves) of the adsorption of fibrinogen on Au, TiO$_2$ and TiO$_2$ + APMS + heparin surfaces in real time. (b) Correlation between ΔD and Δf of the different surfaces from Figure 4.41.

In contrast, the ratio is generally higher for the heparinized surfaces with an increase after approx. $\Delta f = -20$ Hz. This behavior indicates structural changes within the protein layer [88]. The higher dissipation on the heparin-modified surface indicates a lower viscosity of the protein film on this surface. Since a densification of protein films is normally accompanied by conformational changes, it is likely that these changes are less pronounced on the heparin surfaces compared to gold or pure TiO$_2$.

4.3.4
Surface Modification of Titanium for Improvement of the Interfacial Biocompatibility

Doris Klee, Jürgen Böing, and Hartwig Höcker

The chemical vapor deposition (CVD) polymerization of amino-functionalized [2.2]-*para*-cyclophane was carried out for polymer coating and functionalization of titanium surfaces. The generated amino groups were used for covalent immobilization of bioactive substances to stimulate the adhesion and growth of osteoblasts. As bioactive substances the pentapeptide GRGDS and the growth factor bone morphogenetic protein 2 (BMP-2) were chosen. The covalent bonding was achieved by activation with hexamethylene diisocyanate. Each modification step was characterized by X-ray photoelectron spectroscopy (XPS) and contact angle measurements. The covalent bonding of the bioactive substances was proven by radiolabeling and *surface*-MALDI-TOF-MS. *In vitro* biocompatibility tests with primary, human osteoblasts demonstrated the improved cell adhesion and *spreading* on the bioactive modified titanium surfaces.

Introduction
In order to control osteointegration on implants made of titanium material (alloy Ti6Al4V), they have been functionalized with bioactive substances which enhance bone cell adhesion and stimulate their growth. As mentioned, fibronectin and its sequence arginine-glycine-aspargic acid (RGD), from the cell binding domain, have been chosen as cell mediators [89]. Moreover, growth factors such as BMP-2 are attracting increasing interest [90]. For a permanent and strong bonding between the titanium surface and the bone tissue, it is necessary that the active substance be coupled covalently. By this means, desorption and diffusion of the active substance in the tissue is avoided and the activity of the molecule is confined to the interface. Covalent immobilized biofunctional coatings demand appropriate chemical groups on the titanium surface, which is originally very inert. Besides the silanization with 3-aminopropyltriethoxysilane, the generation of the suitable groups can be achieved by CVD polymerization with amino-functionalized [2.2]-*para*-cyclophanes [91,92]. Additionally linker molecules are often used to separate surface and active substance so that the mobility and accessibility to the active substance is improved, while changes in conformation are limited. Hexamethylendiisocyanate (HDI) is one of possible used linkers. The surface modification steps are presented in a scheme in Figure 4.27.

Functionalization of Titanium Surface by CVD Polymerization
The CVD polymerization of amino-substituted *para*-cyclophane (A-PCP) is a deposition without solvent or initiator that simultaneously coats the titanium surface and supplies the functional groups on it in a unique process step. The synthesis of single-substituted 4-amino-[2.2]-*para*-cyclophane (A-PCP) is done in a five-step reaction from *para*-cyclophane [93]. The CVD polymerization of A-PCP on cleaned titanium surfaces based on Ti6Al4V alloy was performed in a laboratory machine based on the Gorham process [95]. First of all, the A-PCP is sublimed under low pressure. It is then pyrolized at 700 °C and 0.2 mbar. Finally A-PCP polymerizes on the substrate

Evaporation   **Pyrolysis**   **Deposition**

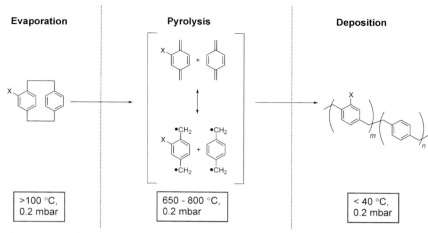

| >100 °C, 0.2 mbar | 650 - 800 °C, 0.2 mbar | < 40 °C, 0.2 mbar |

Figure 4.43 Mechanism of the CVD polymerization of functionalized [2.2]*para*-cyclophanes. Scheme of the Gorham process for the CVD polymerization [35].

at temperatures below 25 °C producing a poly(amino-*p*-xylylen-*co*-poly-*p*-xylylen) (amino-ppx) coating containing primary amino groups, thus enabling the immobilization step afterwards (see Figure 4.43).

The chemical surface composition of the polymer film generated on the titanium was characterized by XPS. Table 4.17 summarizes the measured atomic composition of the titanium surface prior and after CVD with amino-ppx as well as the calculated element composition of the pure amino-ppx. The non-cleaned titanium surface amounts to 49.6 at.% carbon; after cleaning there is still 18/1 at.% carbon [98,99]. The remaining carbon on the surface can be ascribed to adsorbed hydrocarbons and bound alkoxides and carboxylates. The XPS measurements confirm the amino groups from the amino-ppx polymer coating. The fact that the amino-ppx coating

Table 4.17 Element composition of a titanium surface in at.% before and after coating with amino-ppx according to XPS compared to the calculated elemental composition of pure amino-ppx.

Photoline	C_{1s}			Ti_{2p}	O_{1s}		N_{1s}	Rest
E_B (eV)	285.0	286.5	289.1	458.2	530.4	532.2	400.0	–
Structure element	$\underline{C}-C$	$\underline{C}-O$ $\underline{C}-N$	$O-\underline{C}=O$	$Ti\underline{O}_2$	$Ti\underline{O}_2$	$\underline{O}-C$	$\underline{N}-C$	–
Titanium cleaned (measured in at.%)	13.3	3.5	1.3	20.0	41.0	16.4	–	4.5[a]
A-PPX (calculated in at.%)	88.2	5.9	–	–	–	–	5.9	–
Titan/A-PPX (measured in at.%)	85.0	6.8	–	–	–	2.6	5.6	–

[a] (Al, V).

contains oxygen is due to the saturation of the radicals with oxygen from the air during the reaction or when ventilating the CVD machine after the coating process. As titanium is not measured any longer after the CVD, it is concluded that the coating is homogeneous. Depending on the amount of sublimed A-PCP the layer thickness can be controlled. In the reported experiments a layer thickness of nearly 100 nm was reached. The static contact angle of water on the cleaned titanium surfaces was 20°, showing a high hydrophilicity. After the CVD coating with amino-ppx, the contact angle increased markedly to 82°.

Immobilization of Bioactive Substances via Aminofunctionalized Poly (*p*-xylylenes)
The pentapeptide GRGDS was chosen as biological signal as it contains RGD, a sequence recognized by cells. Additionally, BMP-2 was chosen because of its osteoinductive properties and its differentiation enhancement on mesenchymal bone cells [91]. Jennissen et al. accomplished the first investigations of BMP-2 immobilization on metallic implant surfaces [97]. The immobilization of the RGD peptide and BMP-2 on the HDI-activated samples was carried out in sodium hydrogen carbonate buffer at pH = 8.4 (concentrations were 100 and 50 μg mL^{-1} respectively). In order to quantify the amount of RGD peptide bound to the surface, YRGDS was selected as peptide. The n-terminus glycine is exchange by tyrosine (Y) because the aromatic ring of tyrosine can be marked with ^{125}Iod by the Chloramin-T method and hence its radioactivity can be measured. It is supposed that the adsorption of both peptides does not differ significantly and that the results obtained with the immobilized YRGDS would be comparable to those of the coupling of GRGDS under the same conditions. The amount of labeled YRGDS on the reference TCPS was 26 ng cm^{-2}, whereas the adsorption on non-treated and cleaned titanium surfaces was 14 and 10 ng cm^{-2} respectively.

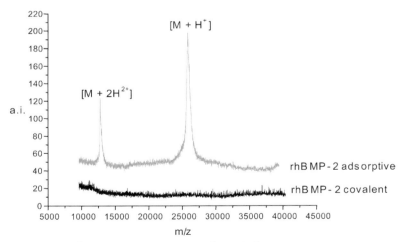

Figure 4.44 Surface-MALDI-TOF-MS spectra of BMP-2 after immobilization on amino-ppx-coated titanium surfaces with and without activation by HDI [38].

The immobilization of BMP-2 on the corresponding surfaces was proven qualitatively by means of XPS because of the changes in surface composition [101]. Surface-MALDI-TOF-MS was used to differentiate between adsorbed and covalently immobilized BMP-2 on the functionalized titanium surfaces after HDI activation. The immobilization on the amino-ppx-functionalized surfaces was carried out with and without HDI activation. It is supposed that during the desorption process with laser energy only the molecules adsorbed physically, and not the chemically bound ones, can be detected [100]. Figure 4.44 shows the surface-MALDI-ToF mass spectra of BMP-2 on titanium surfaces coated with amino-ppx with and without HDI linker. Only the surface without HDI activation shows the molecule peaks of BMP-2 at m/z 13 000 and 26 000. This implies that the protein could not be removed completely from the surface during the washing procedure and that it remains partially adsorbed on it (see Figure 4.44).

The biological response of the modified titanium surfaces was tested *in vitro* in cell experiments with primary human osteoblasts. The vitality of the osteoblasts was characterized by fluorescence staining with Fluoresceindiacetat/Ethidiumbromid. Besides cell vitality, this test also facilitates the examination of the morphology of the adhered cells. Figure 4.45 presents images of stained osteoblasts after 24 h on cleaned and amino-ppx-functionalized titanium surfaces observed with a fluorescence

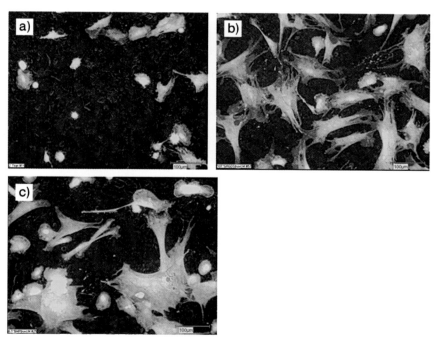

Figure 4.45 Vitality of osteoblasts after 24 h contact: cleaned titanium surface (a), GRGDS covalently immobilized on amino-ppx-coated titanium surface (b), BMP-2 covalently immobilized on amino-ppx-coated titanium surface (c) ×100/×200 [40].

microscope. The vitality of the osteoblasts on titanium surfaces modified covalently with biological substances is substantially improved by the presence of the RGD peptide. Likewise, the covalent immobilization of the growth factor BMP-2 stimulates considerably the spreading of the osteoblasts. Compared to the coupling of RGD peptide, BMP-2 results in an even stronger spreading.

Conclusion

The biocompatibility investigations with primary human osteoblasts on titanium surfaces biofunctionalized with biological cell mediator substances have shown promising results which announce a possible transfer to titanium implants. Particularly, CVD as a solvent-free and single-step process is of interest for its economic advantages. The CVD process described here establishes the feasibility, on requirement, to couple covalently other biologically active molecules on functionalized titanium surfaces.

Acknowledgment

For the performance of the *in vitro* cell experiments the authors thank Dr. C. Niedhart from the Orthopädischen Klinik der RWTH Aachen.

4.3.5
Biofunctionalization of Titanium Substrates Using Maleic Anhydride Copolymer Films

Tilo Pompe, Carsten Werner, and Hartmut Worch

The interaction of an implant surface with the living environment very sensitively determines the performance of the implant. Adsorption and subsequent conformational changes of adsorbed biomolecules, like proteins or glycosaminoglycans, trigger the host response concerning incorporation, inflammation or encapsulation [101]. At the same time it is known that the implant–biomolecule interaction can distinctively be adjusted by changing the physicochemical surface characteristics [102]. By changing functional chemical groups on the implant surface [103], introducing spacer functions or varying the elastic properties of the topmost layer, cell behavior can be affected in respect to adhesion, proliferation and differentiation [104].

An approach was undertaken to modify the surface of flat and microstructured titanium substrates by functional thin polymer films. In previous studies we have already demonstrated that cellular adhesions and cell morphology can be directly influenced by the usage of thin films of different maleic anhydride copolymers [105–107]. Therefore, we made an attempt to adopt this functionalization strategy to titanium surfaces.

Materials and Methods

Titanium (Ti) or titanium 30%/silicon 70% (Ti30Si70) surfaces were prepared by sputtering 100 nm layers on top of clean glass coverslips or on silicon wafers patterned with a topography of micrometer-sized grooves with sizes ranging from

5 to 80 μm. These grooved structures were prepared by a lithographic process with a 10 μm deep etch. By wet chemical surface preparation thin films of different maleic anhydride copolymers (poly(octadecene-*alt*-maleic anhydride) (POMA; MW = 40 000 g mol^{-1}), poly(propene-*alt*-maleic anhydride) (PPMA; MW = 39 000 g mol^{-1}) and poly(ethylene-*alt*-maleic anhydride) (PEMA; MW = 125 000 g mol^{-1}) were covalently attached via aminosilanes to these surfaces [108].

Thin-film preparation was analyzed by X-ray photoelectron spectroscopy (XPS) using an Amicus spectrometer [109].

Protein attachment and displacement experiments were carried out with solutions of rhodamine-labeled fibronectin or bovine serum albumin in PBS buffer (50 μg mL^{-1}). In displacement experiments the samples were subsequently immersed in 500 μg mL^{-1} human serum albumin in PBS buffer for 24 h [110]. In cell culture experiments human umbilical vein endothelial cells were grown for 50 min on top of fibronectin-coated surfaces with endothelial cell growth medium [105].

For fluorescence microscopy, cells were fixed, permeabilized and stained with phalloidin-FITC.

Results

Titanium or Ti30Si70 substrates were functionalized with amine groups with (3-aminopropyl)triethoxysilane from a ethanol/water solution. This preparation resulted in a dense covalently attached silane layer as revealed from XPS with a nitrogen content of approximately 4 at%.

By spin-coating of maleic anhydride copolymer solutions onto the silane-modified substrates thin polymer films were formed, which were covalently attached by a subsequent temperature treatment at 120 °C to form stable imide bonds. The quantification of the XPS analysis is shown in Table 4.18. As the underlying substrate consists of silicon and titanium with an aminosilane overcoat, the detection of silicon, titanium and nitrogen signals indicates a copolymer layer thickness in the range of few nanometers. Furthermore, the small differences in these signals indicate slight variations in the copolymer layer thickness, which are in agreement with previously published results on silicone oxide surfaces [108] and can be attributed to the different molecular weight of the copolymers used. The high-resolution spectra of the C$_{1s}$ peak (Figure 4.46) demonstrated the different characteristics of the copolymer coatings. The peak at 289.4 eV results from the oxygen-associated carbon atoms in the

Table 4.18 XPS of Ti30Si70 surfaces after modification with different maleic anhydride copolymers. The quantitative atomic composition of the analysis elements is presented.

	C (at.%)	O (at.%)	Si (at.%)	N (at.%)	Ti (at.%)
POMA	73.8	17.6	6.8	1.6	0.1
PPMA	58.7	27.4	11.0	2.2	0.7
PEMA	61.5	28.9	7.4	1.8	0.3

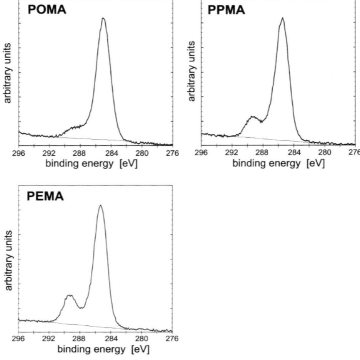

Figure 4.46 High-resolution C_{1s} XPS spectra of Ti30Si70 surfaces after surface modification with different maleic anhydride copolymers. The successful coating of the surfaces with different copolymers is demonstrated by the height of the peak at 289.4 eV from the oxygen-associated carbon atoms of the anhydride ring.

anhydride ring. Hence, its height indicates the relative amount of functional anhydride groups in comparison the other carbon atoms from the polymer chain. Thus, the different amounts indicate the characteristics of the three copolymers, originating from the different size ratio of the anhydride group and the alkyl comonomer with the small size of anhydride in comparison to the long octadecence alkyl chain in POMA and a similar size for PPMA and PEMA due to their much shorter propene or ethylene comonomers, respectively.

The functionality of the copolymer coatings was tested in protein displacement experiments, where substrates with initially attached labeled bovine serum albumin were exposed to unlabeled human serum albumin over 24 h. Figure 4.47 shows the displacement characteristics of four different surfaces. The anhydride group of the maleic anhydride copolymer provides the possibility of switching between covalent attachment via binding of free amine groups of proteins to the anhydride (P(x)MA-c) or physisorption of proteins to surfaces with anhydride groups already hydrolyzed into carboxylic acid groups (P(x)MA-p). Additionally, the different comonomers of the maleic anhydride copolymers determine the amount of polar and hydrophobic

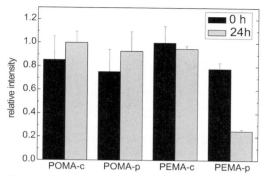

Figure 4.47 Protein exchange of fluorescently labeled bovine serum albumin by human serum albumin over 24 h. The relative intensity for covalently attached (P(x)MA-c) and physisorbed (P(x) MA-p) is plotted for POMA and PEMA surfaces.

interactions between the surfaces and the proteins with POMA providing the longest alkyl comonomer and, hence, the strongest hydrophobic interaction. Accordingly, the displacement probability of the labeled bovine serum albumin is quite low on POMA surfaces for covalent and noncovalent attached protein. Again, a low displacement is observed for covalent attachment on the more hydrophilic PEMA. However, for the hydrolyzed version of PEMA a strong displacement is found due to the much weaker interaction of albumin with the more polar and hydrophilic PEMA surface.

The different attachment strength of proteins to the surfaces could also be shown in cell culture experiments. Pre-coated labeled fibronectin was adsorbed onto the

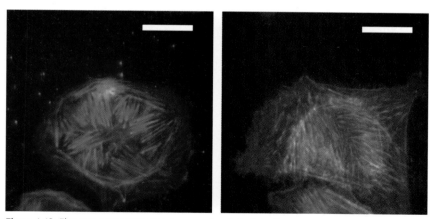

Figure 4.48 Fluorescence microscopy images of endothelial cells grown for 50 min on functionalized titanium surfaces. The cells (visualized by staining the actin cytoskeleton: green) can reorganize labeled fibronectin (red) into fibrillar structures on surfaces with weak fibronectin–substrate interaction (right). On substrates with a strong covalent attachment (left) cells adhere without reorganizing fibronectin. Scale bar: 25 μm.

surfaces. As this protein is a ligand for cell surface adhesion receptors (integrins), endothelial cells could adhere to the substrates. Depending on the fibronectin–substrate attachment strength, the cells can reorganize fibronectin into fibrils. The formation of fibrillar fibronectin networks is a process occurring *in vivo* during tissue generation. Figure 4.48 shows two examples of an endothelial cell grown on PEMA surface with covalent attachment and no fibronectin reorganization (left) and on a hydrolyzed PEMA surface with physisorbed fibronectin and extensive reorganization of fibronectin into fibrils (right).

With the two examples of fibronectin reorganization and albumin displacement it could be nicely demonstrated that titanium surfaces can be functionalized with polymer films in order to modulate their surface physicochemistry. In this way the interaction strength, conformation and mobility of surface-bound biomolecules, e.g. proteins, can adjusted to control cell behavior in respect to adhesion, proliferation and differentiation.

References

1 Scherrer, P. (1918) Bestimmung der Größe und der inneren Struktur von Kolloidteilchen mittels Röntgenstrahlen, *Nachrichten von der Gesellschaft der Wissenschaften zu Göttingen*, **2**, 98–100.

2 Yamaguchi, O., Tomihisa, D., Ogiso, N. and Shimizu, K. (1986) Crystallization of monoclinic $2TiO_2 \cdot 5Nb_2O_5$, *J. Am. Ceram. Soc.*, **69** (7), C150–C151.

3 Geiculescu, A.C. and Spencer, H.G. (1999) Effect of oxygen environment on the decomposition and crystallization of an aqueous sol–gel-derived zirconium acetate gel, *J. Sol–gel Sci. Technol.*, **14**, 257–272.

4 Heidenau, F., Scheler, S., Detsch, R. and Ziegler, G. (2006) Micro structuring of implant surfaces by particle filled sol–gel coatings and cell reaction *in vitro*, *BIOmaterialien*, **7** (S1), 72.

5 Schmidt, H., Heidenau, F. and Ziegler, G. (1999) Sol–gel derived and supercritical dried oxide ceramics, *Adv. Sci. Technol.*, **15**, 143–150.

6 Eisenbarth, E., Velten, D., Schenk-Meuser, K., Linez, P., Biehl, V., Duschner, H., Breme, J. and Hildebrand, H. (2002) Interactions between cells and titanium surfaces, *Biomol. Eng.*, **19**, 243–249.

7 Hallab, N.J., Jacobs, J.J., Skipor, A., Black, J., Mikecz, K. and Galante, J.O. (2000) Systemic metal–protein binding associated with total joint replacement arthroplasty, *J. Biomed. Mater. Res.*, **49**, 353–361.

8 Schaffer, A.W., Pilger, A., Engelhardt, C., Zweymueller, K. and Ruediger, H.W. (1999) Increased blood cobalt and chromium after total hip replacement, *Clin. Toxicol.*, **37**, 839–844.

9 Hoffmann, B., Kokott, A., Shafranska, O., Detsch, R., Winter, S., Eisenbart, E., Peters, K., Breme, J., Kirkpatrick, C.J. and Ziegler, G. (2005) Einfluss einer TiO_2-Beschichtung auf Bio-kompatibilität, Korrosions- und Auslaugverhalten verschiedener Implantatlegierungen, *Biomed. Technol.*, **50**, 320–329.

10 Browne, M. and Gregson, P.J. (1994) Surface modification of titanium alloy implants, *Biomaterials*, **15**, 894–898.

11 Velten, D., Biehl, V., Aubertin, F., Valeske, B., Possart, W. and Breme, J. (2002) Preparation of TiO_2 layers on cp-Ti and Ti6Al4V by thermal and anodic

oxidation and by sol–gel coating techniques and their characterization, *J. Biomed. Mater. Res.*, **59** (1), 18–28.

12 Göbel, M., Sunderkötter, J.D., Mircea, D.I., Jenett, H. and Stroosnijder, M.F. (2000) Study of the high-temperature oxidation behaviour of Ti and Ti4Nb with SNMS using tracers, *Surf. Interf. Anal.*, **29**, 321–324.

13 Park, Y. and Butt, D. (1999) Composition dependence of the kinetics and mechanisms of thermal oxidation of titanium-tantalum alloys, *Oxid. Met.*, **51**, 383–402.

14 Ask, M., Rolander, U., Lausmaa, J. and Kasemo, B. (1990) Microstructure and morphology of surface oxide films on Ti-6Al-4V, *J. Mater. Res. Soc.*, **5**, 1662–1667.

15 Serruys, Y., Sakout, T. and Gorse, D. (1993) Anodic oxidation of titanium in 1M H_2SO_4, studied by Rutherford backscattering, *Surf. Sci.*, **282**, 279–287.

16 National Bureau of Standards (US) Monograph 25 (20), (1983) 115 (JCPDS 34–415).

17 Wadsley, A.D. (1961) Mixed oxides of titanium and niobium: II. The crystal structures of the dimorphic forms $Ti_2Nb_{10}O_{29}$, *Acta Crystallogr.*, **14**, 664–670.

18 Noguchi, T. and Mizuno, M. (1968) Phase changes in the ZrO_2-TiO_2 system, *Bull. Chem. Soc. Jpn.*, **41**, 2895–2899.

19 Jongejan, A. and Wilkins, A.L. (1969) Phase relationships in the high-lime part of the system CaO-Nb_2O_5-SiO_2. *J. Less-Common Metals*, **19**, 203–208.

20 Kittaka, S., Matsuno, K. and Takahara, S. (1997) Transformation of ultrafine titanium dioxide particles from rutile to anatase at negatively charged colloid surfaces, *J. Solid State Chem.*, **132**, 447–450.

21 Hums, E. and Range, K.J. (1990) Investigation of the kinetics of the anatase–rutile transformation on denox-catalyzers for V_2O_5-WO_3-TiO_2 (anatase) and V_2O_5-MoO_3-TiO_2 (anatase), *Ber. Bunsenges. Phys. Chem.*, **94** (6), 691–699.

22 Depero, L.E., Sangaletti, L., Allieri, B., Pioselli, F., Casale, C. and Notaro, M. (1998) Microstructural properties of Ta-doped TiO_2 powders obtained by laser pyrolysis, *Mater. Sci. Forum*, **278–281**, 654–659.

23 Lausmaa, J., Kasemo, B. and Mattson, H. (1990) Surface spectroscopic characterization of titanium implant materials, *Appl. Surf. Sci.*, **44**, 133–146.

24 Zitter, H. and Plenk, J. (1987) The electrochemical behavior of metallic implant materials as an indicator of their biocompatibility, *J. Biomed. Mater. Res.*, **21**, 881–896.

25 Helsen, J.A. and Breme, H.J. (1998) Metals as Biomaterials, John Wiley, New York.

26 Williams, D.F. (1987) Tissue–biomaterials interactions, *J. Mater. Sci.*, **22**, 3421–3445.

27 Ratner, B.D. and Hoffmann, A.S. (2004) Physicochemical surface modification of materials used in medicine, in Bio-materials Science: An Introduction to Materials in Medicine, 2nd edn, (eds B.D. Ratner, A.S. Hoffman, F.J. Schoen and J. E. Lemons), Elsevier, Amsterdam, pp. 201.

28 Jansen, J.A. and von Reum, A.F. (2004) Textured and porous materials, in Biomaterials Science: An Introduction to Materials in Medicine, 2nd edn, (eds B.D. Ratner, A.S. Hoffman, F.J. Schoen, and J.E. Lemons), Elsevier, Amsterdam, pp. 218.

29 Hoffman, A.S. and Hubbell, J.A. (2004) Surface-immobilized biomolecules, in Biomaterials Science: An Introduction to Materials in Medicine, 2nd edn, (eds B.D. Ratner, A.S. Hoffman, F.J. Schoen and J.E. Lemons), Elsevier, Amsterdam, pp. 225.

30 Vallet-Regí, M. and González-Calbet, J. M. (2004) Calcium phosphates as substitution of bone tissues, *Prog. Solid State Chem.*, **32**, 1–31.

31 Landi, E., Tampieri, A., Celotti, G., Langenati, R., Sandri, M. and Sprio, S. (2005) Nucleation of biomimetic apatite in synthetic body fluid: dense and porous scaffold development, *Biomaterials*, **26**, 2835–2845.

32 Barrere, F., van Blitterswijk, C.A., de Groot, K. and Layrolle, P. (2002) Influence of ionic strength and carbonate on the Ca-P coating formation from SBF × 5 solution, *Biomaterials*, **23**, 1921–1930.

33 Rambo, C.R., Müller, F.A., Müller, L., Sieber, H., Hofmann, I. and Greil, P. (2006) Biomimetic apatite coating on biomorphous alumina scaffolds, *Mater. Sci. Eng.*, **C26**, 92–99.

34 Müller, F.A., Müller, L., Hofmann, I., Greil, P., Wenzel, M.M. and Staudenmaier, R. (2006) Cellulosebased scaffold materials for cartilage tissue engineering, *Biomaterials*, **27**, 3955–3963.

35 Jonasova, L., Müller, F.A., Helebrant, A., Strnad, J. and Greil, P. (2004) Biomimetic apatite formation on chemically treated titanium, *Biomaterials*, **25**, 1187–1194.

36 Müller, L. and Müller, F.A. (2006) Preparation of SBF with different HCO_3^- content and its influence on the composition of biomimetic apatites, *Acta Biomater.*, **2**, 181–189.

37 Zhu, P.X., Masuda, Y., Yonezawa, T. and Koumoto, K. (2003) Investigation of apatite deposition onto charged surface in aqueous solutions using a quarz-crystal microbalance, *J. Am. Ceram. Soc.*, **86**, 782–790.

38 Ito, A., Maekawa, K., Tsutsumi, S., Ikazaki, F. and Tateishi, T. (1997) Solubility product of OH-carbonated hydroxyapatite, *J. Biomed. Mater. Res.*, **36**, 522–528.

39 Nancollas, G.H. (1989) in Biomineralization, (eds S. Mann, J. Webb, and R.J.P. Williams), VCH Verlag, Weinheim, pp. 159.

40 Tang, R., Henneman, Z.J. and Nancollas, G.H. (2003) Constant composition kinetics study of carbonated apatite dissolution, *J. Cryst. Growth*, **249**, 614–624.

41 Davies, J.W. (1962) Ion Association, Butterworth Press, London.

42 Siripphannon, P., Kameshina, Y., Yasumori, A., Okada, K. and Hayashi, S. (2002) Comparative study of the formation of hydroxyapatite in simulated body fluid under static and flowing system, *J. Biomed. Mater. Res.*, **60**, 175–185.

43 Hench, L.L. and Latorre, G.P. (1992) Reaction kinetics of bioactive ceramics: IV. Effect of glass and solution composition, in Biomaterials 5, (eds T. Yamamuro, T. Kokubo and T. Nakamura), Kobonshi Kankokai, Kyoto, pp. 67.

44 Elliot, J.C. (1994) Structure and Chemistry of the Apatites and Other Calcium Orthophosphates, Elsevier Press, Amsterdam.

45 Jonasova, L., Müller, F.A., Helebrant, A., Strnad, J. and Greil, P. (2002) Hydroxyapatite formation on alkali treated titanium with different content of Na^+ in the surface layer, *Biomaterials*, **23**, 3095–3101.

46 Tadic, D., Peters, F. and Epple, M. (2002) Continuous synthesis of amorphous carbonated apatites, *Biomaterials*, **23**, 2553–2559.

47 Danilchenko, S.N., Kukharenko, O.G., Moseke, C., Protsenko, I.Y., Sukhodub, L.F. and Sulkio-Cleff, B. (2002) Determination of the bone mineral crystallite size and lattice strain from diffraction line broadening, *Cryst. Res. Technol.*, **37**, 1234–1240.

48 Weiner, S. and Wagner, H.D. (1998) The material bone: structure–mechanical function relation, *Annu. Rev. Mater. Sci.*, **28**, 271–298.

49 Müller, F.A., Müller, L., Caillard, D. and Conforto, E. (2007) Preferred growth orientation of biomimetic apatite crystals, *J. Cryst. Growth*, **304**, 464–471.

50 Stoch, A., Jastrzebski, W., Brozek, A., Stoch, J., Szatraniec, J., Trybalska, B. and Knita, G. (2000) FTIR absorption–reflection study of biomimetic growth of phosphates on titanium implants, *J. Mol. Struct.*, **555**, 375–382.

51 Koutsopoulos, S. (2002) Synthesis and characterization of hydroxyapatite crystals: a review study on the analytical methods, *J. Biomed. Mater. Res.*, **62**, 600–612.

52 Simpson, L.J. (1998) Electrochemically generated CaCO₃ on iron studied with FTIR and Raman spectroscopy, *Electrochim. Acta*, **43**, 2543–2547.

53 Krajewski, A., Mazzocchi, M., Buldini, P.L., Ravaglioli, A., Tinti, A., Taddei, P. and Fagnano, C. (2005) Synthesis of carbonated hydroxyapatites: efficiency of the substitution and critical evaluation of analytical methods, *J. Mol. Struct.*, **744–747**, 221–228.

54 Wilson, R.M., Elliot, J.C., Dowkner, S.E.P. and Rodriguez-Lorenzo, P. (2005) Rietveld refinements and spectroscopic studies of the structure of Ca-deficient apatite, *Biomaterials*, **26**, 1317–1327.

55 Slosarczyk, A., Paszkiewicz, Z. and Paluszkiewicz, C. (2005) FTIR and XRD evaluation of carbonated hydroxyapatite powders synthesized by wet method, *J. Mol. Struct.*, **744–747**, 653–656.

56 Pasteris, J.D., Wopenka, B., Freeman, J., Rogers, K., Valsami-Jones, E., van der Houwen, J.A.M. and Silva, M.J. (2004) Lack of OH in nanocrystaline apatite as a function of degree of atomic order: implication for bone and biomaterials, *Biomaterials*, **25**, 229–238.

57 Landi, E., Tampieri, A., Celotti, G., Vichi, L. and Sandri, M. (2004) Influence of synthesis and sintering parameters on the characteristics of carbonate apatite, *Biomaterials*, **25**, 1763–1770.

58 Wilson, R.M., Elliott, J.C., Dowker, S.E.P. and Smith, R.I. (2004) Rietveld structure refinement of precipitated carbonate apatite using neutron diffraction data, *Biomaterials*, **25**, 2205–2213.

59 Fleet, M. and Liu, X.Y. (2004) Location of type B carbonate ion in type A-B carbonate apatite synthesized at high pressure, *J. Solid State Chem.*, **177**, 3174–3182.

60 LeGeros, R.Z. (1991) Calcium Phosphates in Oral Biology and Medicine, Karger, Basel.

61 Matthew, J.K., Rakovan, J. and Hughes, J.M. (2002) Phosphates: geochemical, geobiological and materials importance, *Rev. Mineral. Geochem.*, **48**, 26.

62 ElFeki, H., Savariault, J.M., Ben Salah, A. and Jemal, M. (2000) Sodium and carbonate distribution in substituted calcium hydroxyapatite, *Solid State Sci.*, **2**, 577–586.

63 El-Ghannam, A.R., Ducheyne, P., Risbud, M., Adams, C.S., Shaprio, I.M., Csatner, D., Golledge, S. and Composto, R.J. (2004) Model surfaces engineered with nanoscale roughness and RGD tripeptides promote osteoblast activity, *J. Biomed. Mater. Res.*, **68A**, 615–627.

64 Mann, B.K. and West, J.L. (2002) Cell adhesion peptides alter smooth muscle cell adhesion, proliferation, migration, and matrix protein synthesis on modified surfaces and in polymer scaffolds, *J. Biomed. Mater. Res.*, **60**, 86–93.

65 Hersel, U., Dahmen, C. and Kessler, H. (2003) RGD modified polymers: biomaterials for stimulated cell adhesion and beyond, *Biomaterials*, **24**, 4385–4415.

66 Peramo, A., Albritton, A. and Matthews, G. (2006) Deposition of patterned glysosaminoglycans on silanized glass surfaces, *Langmuir*, **22**, 3228–3234.

67 Smith, J.T., Elkin, J.T. and Reichert, W.M. (2006) Directed cell migration on

fibronectin gradients: effect of gradient slope, *Exp. Cell Res.*, **312**, 2424–2432.

68 Scheideler, L., Rupp, F., Wendel, H.P., Sathe, S. and Geis-Gerstorfer, J. (2007) Photocoupling of fibronectin to titanium surfaces influences keratinocyte adhesion, pellicle formation and thrombogenicity. Dental Mater., **23**, 469–478.

69 Puleo, D.A., Kissling, R.A. and Sheu, M.-S. (2002) A technique to immobilize bioactive proteins, including bone morphogenetic protein-4 (BMP-4), on titanium alloy, *Biomaterials*, **23**, 2079–2087.

70 Xiao, S.J., Textor, M., Spencer, N.D., Wieland, M., Keller, B. and Sigrist, H. (1997) Immobilization of the cell-adhesive peptide Arg-Gly-Asp-Cys (RGDC) on titanium surfaces by covalent chemical attachment, *J. Mater. Sci. Mater. Med.*, **8**, 867–872.

71 Porté-Durrieu, M.C., Labrugère, C., Villars, F., Lefebvr, F., Dutoya, S., Guette, A., Bordenave, L. and Baquey, C. (1999) Development of RGD peptides grafted onto silica surfaces: XPS characterization and human endothelial cell interactions, *J. Biomed. Mater. Res.*, **46**, 368–375.

72 Schuler, M., Owen, G.R., Hamilton, D.W., de Wild, M., Textor, M., Brunette, D.M. and Tosatti, S.G.P. (2006) Biomimetic modification of titanium dental model surfaces using the RGDSP-peptide sequence: a cell morphology study, *Biomaterials*, **27**, 4003–4015.

73 Müller, R., Abke, J., Schnell, E., Scharnweber, D., Kujat, R., Englert, C., Taheri, D., Nerlich, M. and Angele, P. (2006) Influence of surface pretreatment of titanium- and cobalt-based biomaterials on covalent immobilization of fibrillar collagen, *Biomaterials*, **27**, 4059–4068.

74 Müller, R., Abke, J., Macionczyk, F., Gbureck, U., Ruszczak, Z., Mehrl, R.,

Kujat, R., Englert, C., Nerlich, M. and Angele, P. (2005) Surface engineering of stainless steel materials by covalent collagen immobilization to improve implant biocompatibility, *Biomaterials*, **26**, 6962–6972.

75 Cook, A.D., Pajvani, U.B., Hrkach, J.S., Canninzzaro, S.M. and Langer, R. (1997) Colorimetric analysis of surface reactive amino groups on poly(lactic acid-co-lysine)–poly(lactic acid) blends, *Biomaterials*, **18**, 1417–1424.

76 Geißler, U., Hempel, U., Wolf, C., Scharnweber, D., Worch, H. and Wenzel, K.W. (2000) Collagen type I-coating of Ti6Al4V promotes adhesion of osteoblasts, *J. Biomed. Mater. Res.*, **51**, 752–760.

77 Angele, P., Abke, J., Kujat, R., Faltermeier, H., Schumann, D., Nerlich, M., Kinner, B., Englert, C., Ruszczak, Z., Mehrl, R. and Müller, R. (2004) Influence of different collagen species on physico-chemical properties of crosslinked collagen matrices, *Biomaterials*, **25**, 2831–2841.

78 Shafi, K.V.P.M., Ulman, A., Yan, X., Yang, N.-L., Himmelhaus, M. and Grunze, M. (2001) Sonochemical preparation of silane-coated titania particles, *Langmuir*, **17**, 1726–1730.

79 Moore, S. and Stein, W.H. (1948) Photometric ninhydrin method for use in chromatography of amino acids, *J. Biol. Chem.*, **176**, 367–388. Moore, S. and Stein, W.H. (1954) A modified ninhydrin reagent for the photometric determination of amino acids and related compounds, *J. Biol. Chem.*, **211**, 907–913. Moore, S. (1968) Amino acid analysis: aqueous dimethyl sulfoxid as solvent for the ninhydrin reaction, *J. Biol. Chem.*, **243**, 6281–6283.

80 Smith, P.K., Mallia, A.K. and Hermannson, G.T. (1980) Colorimetric method for the assay of heparin content in immobilized heparin preparations, *Anal. Biochem.*, **109**, 466–473.

81 Hall, R. and Malia, R.G. (eds). (1984) Medical Laboratory Haemotology, 1st edn, Butterworths, London, pp. 629.

82 Puleo, D.A. (1997) Retention of enzymatic activity immobilized on silanized Co-Cr-Mo and Ti-6Al-4V, *J. Biomed. Mater. Res.*, **37**, 222–228.

83 Kang, I.K., Kwon, O.H., Lee, Y.M. and Sung, Y.K. (1996) Preparation and surface characterization of functional and heparin-immobilized poly-urethanes by plasma glow discharge, *Biomaterials*, **17**, 841–847.

84 Park, K.D., Kim, W.G., Jacobs, H., Okano, D. and Kim, S.W. (1992) Blood compatibility of SPUU-PEO-heparin grafted copolymers, *J. Biomed. Mater. Res.*, **26**, 739–756.

85 Sanchez, J., Elgue, G., Riesenfeld, J. and Olsson, P. (1997) Inhibition of the plasma contact activation system of immobilized heparin: relation to surface density of functional anti-thrombin binding sites, *J. Biomed. Mater. Res.*, **37**, 37–42.

86 Thull, R. (2002) Physicochemical principles of tissue material inter-actions, *Biomol. Eng.*, **19**, 42–50.

87 Höök, F. (1997) Development of a novel QCM technique for protein adsorption studies. Dissertation, Chalmers University of Technology and Göteborg University.

88 Höök, F., Rodahl, M., Brzezinski, P. and Kasemo, B. (1998) Energy dissipation kinetics for protein and antibody–antigen adsorption under shear oscillation on a quartz crystal micro-balance, *Langmuir*, **14**, 729–734.

89 Pierschbacher, M.D., Hayman, E.G. and Ruoslahti, E. (1983) *Proc. Nat. Acad. Sci. USA*, **80**, 1224.

90 Weimann, E. and Kiess, W. (1995) Wachstumsfaktoren: Grundlagen und klinische Anwendung, 2nd edn, Schattauer Verlag, Stuttgart, New York, pp. 36.

91 Lahann, J., Klee, D., Thelen, H., Bienert, H., Vorwerk, D. and Höcker, H.

(1999) *J. Mater. Sci. Mater. Med.*, **10**, 443.

92 Klee, D., Lahann, J. and Plüster, W. (2002) Dünne Beschichtungen auf Biomaterialien, in: Medizintechnik mit biokompatiblen Wirkstoffen und Verfahren (ed. S.E. Wintermantel, S.-W. Ha), Springer,Berlin, pp. 347.

93 Waters, J.F., Sutter, J.K., Meador, M.A.B., Baldwin, L.J. and Meador, M.A. (1991) *J. Polym. Sci. Part A*, **26**, 1917.

94 Lahann, J., Klee, D. and Höcker, H. (1998) Chemical vapour deposition polymerization of substituted [2.2] paracyclophanes, *Macromol. Rapid Commun.*, **19**, 441–444.

95 Klee, D., Weiss, N. and Lahann, J. (2004) Modern Cyclophane Chemistry (eds R. Gleiter and H. Hopf), Wiley-VCH, Weinheim.

96 Kübler, N.R., Reuther, J.F., Faller, G., Kirchner, T., Ruppert, R. and Sebald, W. (1998) *Int. J. Oral Maxillofac. Surg.*, **27**, 305.

97 Jennissen, H.P., Zumbrink, T., Chatzinikolaidou, M. and Steppuhn, J. (1999) *Mat.-wiss. u. Werkstofftech.*, **30**, 838.

98 Klee, D., Böing, J. and Höcker, H. (2004) Surface Modification of titanium for improvement of the interfacial biocompatibility, *Materialwissenschaft und Werkstofftechnik*, 186–191.

99 Böing, J. (2003) PhD thesis, RWTH Aachen.

100 McLean, K.M., McArthur, S.L., Chatelier, R.C., Kingshott, P. and Griesser, H.J. (2000) *Colloid Surf. B*, **17**, 23.

101 Wilson, C.J., Clegg, R.E., Leavesley, D.I. and Pearcy, M.J. (2005) Mediation of biomaterial–cell interactions by adsorbed proteins: a review, *Tissue Eng.*, **11**, 1–18.

102 Shin, H., Jo, S. and Mikos, A.G. (2003) Biomimetic materials for tissue engineering, *Biomaterials*, **24**, 4353–4364.

103 Keselowsky, B.G., Collard, D.M. and Garcia, A.J. (2005) Integrin binding

specificity regulates biomaterial surface chemistry effects on cell differentiation, *Proc. Natl. Acad. Sci. USA*, **102**, 5953–5957.

104 Engler, A.J., Sen, S., Sweeney, H.L. and Discher, D.E. (2006) Matrix elasticity directs stem cell lineage specification, *Cell*, **126**, 677–689.

105 Pompe, T., Kobe, F., Salchert, K., Jorgensen, B., Oswald, J. and Werner, C. (2003) Fibronectin anchorage to polymer substrates controls the initial phase of endothelial cell adhesion, *J. Biomed. Mater. Res.*, **67A**, 647–657.

106 Pompe, T., Markowski, M. and Werner, C. (2004) Modulated fibronectin anchorage at polymer substrates controls angiogenesis, *Tissue Eng.*, **10**, 841–848.

107 Pompe, T., Keller, K., Mitdank, C. and Werner, C. (2005) Fibronectin fibril pattern displays the force balance of cell–matrix adhesion, *Eur. Biophys. J.*, **34**, 1049–1056.

108 Pompe, T., Zschoche, S., Herold, N., Salchert, K., Gouzy, M.F., Sperling, C. and Werner, C. (2003) Maleic anhydride copolymers: a versatile platform for molecular biosurface engineering, *Biomacromolecules*, **4**, 1072–1079.

109 Pompe, T., Renner, L., Grimmer, M., Herold, N. and Werner, C. (2005) Functional films of maleic anhydride copolymers under physiological conditions, *Macromol. Biosci.*, **5**, 890–895.

110 Renner, L., Pompe, T., Salchert, K. and Werner, C. (2004) Dynamic alterations of fibronectin layers on copolymer substrates with graded physicochemical characteristics, *Langmuir*, **20**, 2928–2933.

II

Physical and Physicochemical Surface Characterization

1

Introduction

Roger Thull

Some materials in implants, in particular spontaneously passivating metals of groups 4b and 5b of the periodic table of elements, are virtually free of reactions with recognizing cells of the immunological system. This and some other physicochemical properties are important sources of biocompatibility. Refractory metals such as titanium, zirconium and their alloys as well as some ceramic materials used for implants characterized by very low disintegration rates prompt and justify the question as to the details of how "physicochemical communication" between material surfaces and the extracellular matrix actually occurs.

The interaction between the surface of an implant and the body electrolyte begins with the adsorption of charged ions and polarizable molecules, as water molecules or biological macromolecules. Through forces of attraction and repulsion and the exchange of charge carriers as electrons for metals or electrons and holes for semiconducting coatings, reactive – reversible or irreversible – structural changes of adsorbed substances can take place. In particular in organic macromolecules intramolecular and intermolecular bonds together with oxygen bridge bonds may break down, giving rise to structural and/or conformational changes. The degree to which this process is reversible will depend upon – for example in the case of proteins – whether the peptide chains are destroyed, or whether disulfide bridges have detached form cysteinyl residues.

Conformational changes may arise as a result of an exchange of charge carriers between the surface of the biomaterial and the biological macromolecules. However, changes can also be initiated by high field strengths, such as occur as a consequence of the presence of local elements, in particular semiconductive or non-conductive surfaces layers.

A conception for the actual changes of the protein molecule during the adsorption process is up to now not available. But it is imaginable that the changes could be similar to those occurring if the pH value of the surrounding electrolyte is changing. The adsorption of hydrogen ions or hydroxyl groups leads to changed potential gradients at the surface of the protein comparable with the potential gradient within the biomaterial/electrolyte interface.

Metallic Biomaterial Interfaces. Edited by J. Breme, C. J. Kirkpatrick, and R. Thull
Copyright © 2008 WILEY-VCH Verlag GmbH & Co. KGaA, Weinheim
ISBN: 978-3-527-31860-5

The equilibrium state of the material surface in the biological environment is characterized by identical rates of adsorption and desorption of ions and constituents of, for instance, the extracellular matrix. Overlapping with the adjustment of equilibrium is the beginning of the interaction between tissue cells and the surface of the material. Cell adhesion involves contact and spreading of cells over the surface, followed by the subsequent differentiation and growth of cells. Cells attach to proteins adsorbed on the surface through specific transmembrane adhesion molecules and to specific sites on these proteins.

A criterion for the suitability of materials as a biomaterial is the physicochemical reactivity of the surface. When a biomaterial (solid) and the extracellular fluid (liquid) come in contact with each other, an interface between them occurs. The region that includes the surface and the immediate adjacent anisotropic boundary is called the interface. The molecules making up a surface are different from those of the bulk phase because they are not surrounded by bulk-phase molecules. The bonding energy of the surface molecule is less than the bonding energy associated with a bulk-phase molecule, and the energy of the surface molecule is therefore higher than that of molecules in the bulk phase.

The following investigations and results are intended to characterize the interface between body fluids and material surfaces, in particular the interface between titanium and physiological saline solutions containing or not containing proteins.

2
Surface Topology

2.1
Mechanical Characterization

Ulrich Beck and Regina Lange

DIN Roughness Parameters

One of the oldest and most common surface topography characterization methods is the identification of characteristic surface profile lines and then the analysis of these lines from the standpoint of their standardized parameters. Profile lines can be identified using the following methods:

- Mechanical tracing using a stylus or other implement.
- Optical scanning using a laser interferometer or other suitable device.
- AFM (suitable for slight roughness only).

The ISO 4287 and 4288 standards that form the basis for surface profile evaluations are mainly based on technical surfaces, and in industrial manufacturing settings allow for rapid characterization of surface topography and valid surface parameter comparisons. However, the question arises as to whether perhaps surface profile characteristics other than average roughness (R_a) should also be used to evaluate the biofunctionality of rough material surfaces and investigate the effect of rough material surfaces on biosystems. The parameter R_a is one of the most common standardized surface parameters and is also used to characterize medical implant surfaces. Inasmuch as average roughness is an integral parameter that is only computed from the roughness profile obtained by filtering of the measured primary profile, this parameter can only provide a rough characterization of surface properties and geometry. The filtering cutoff wavelength is determined by the relevant sampling length, which means that the sampling length selected indirectly affects the R_a value. This scenario clearly shows that it takes a large number of parameters to describe surface characteristics. In addition to average roughness, there is a number of other standardized amplitude and other parameters that can be computed on the basis of the primary and roughness profiles and that to varying extents are suitable for characterization of rough implant

Metallic Biomaterial Interfaces. Edited by J. Breme, C. J. Kirkpatrick, and R. Thull
Copyright © 2008 WILEY-VCH Verlag GmbH & Co. KGaA, Weinheim
ISBN: 978-3-527-31860-5

surfaces. For example, in searching for alternative parameters we used a simple Fourier transformation which yielded good structural feature data. In our view, descriptive functions such as fractal parameters are more suitable than simple roughness parameters for describing the impact of surface on the biofunctionality of complex profiles [1].

This principle is illustrated below using a broad range of modified titanium surfaces as examples in which surfaces with widely varying roughness and structures were obtained using an extremely broad range of processes such as polishing, machining, irradiation with various materials, etching and plasma spraying (see Chapter 3 of Part I). The profile data were registered via tracing (Hommel tester, 1500 mm tracing length, $0.5\,\mathrm{mm\,s^{-1}}$ scan rate) (Figure 2.1).

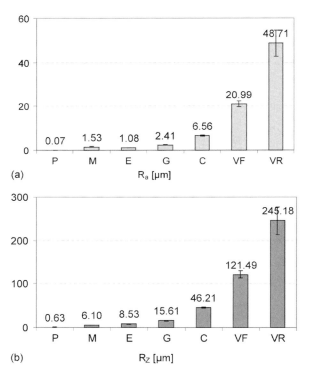

Figure 2.1 DIN roughness parameters R_a (a) and R_z (b) for various types of titanium surface modifications (tracing profiles measured with Hommel tester, 1500 mm tracing length, $0.5\,\mathrm{mm\,s^{-1}}$ scan rate): P, polished; M, machined; E, etched; G, glass bead blasted; C, corundum blasted; VF, vacuum plasma sprayed fine; VR, vacuum plasma sprayed rough.

Fractal Characterization

The concept of self-similarity and (hence the fractal dimension itself) often comes up in connection with rough surface characterization, whereby it is presumed that both natural and technical rough surfaces have self-similar properties and can be regarded as random fractals [2–4].

As a descriptive value, the fractal dimension for self-similarity of a rough surface ranges from a value of 2.0 (even) for a very smooth and even surface, and then approaches 3.0 (spatial) for an extremely roughened surface. An extremely broad range of methods is available for measuring the fractal dimension of a surface. As with a curved or rough surface, the fractal dimension of a bent (profile) line can also be identified, although in such cases the value must range from 1.0 (straight line) to 2.0 (extremely curved and ramified line) [5].

The following equation for computing structure function and the Hurst coefficient H can be used to measure the fractal dimension of a profile line:

$$S(\sigma) = \langle \Delta h^2(\sigma) \rangle = K\sigma^{2H} \quad \text{with} \quad \Delta h(\sigma) = h(x+\sigma) - h(x) \tag{2.1}$$

$S(\sigma)$ describes the mean value for a fixed distance σ for all squared height differences $\Delta h(\sigma)$ within the height profile at a distance of σ. If this structure function is consistent with a power law as described above (i.e. in a double logarithmic graph log $(S) = f(\log(\sigma))$ a straight line results) fractality is deemed to exist and the fractal parameters H (Hurst coefficient) and K can be determined [6–9]. In Figure 2.2 this procedure is demonstrated for a vacuum plasma sprayed sample.

The fractal dimension D_p is based on the following:

$$D_p = 2 - H \tag{2.2}$$

and the topothesy L from

$$K = L^{2-2H} \tag{2.3}$$

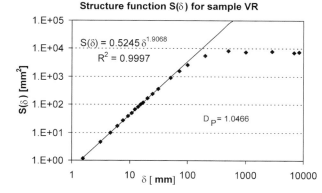

Structure function S(δ) for sample VR

$S(\delta) = 0.5245\,\delta^{1.9068}$
$R^2 = 0.9997$

$D_p = 1.0466$

Figure 2.2 Example for the determining of the fractal dimension D_p using a two-dimensional structure function $S(\sigma)$ calculated from the tracing profile (Hommel tester, 1500 mm tracing length, 0.5 mm s^{-1} scan rate) for a vacuum plasma sprayed sample VR.

The structure function is closely related to the autocorrelation function in telecommunications and can be also be computed three dimensionally (e.g. from AFM measurement data). In such cases the fractal dimension of a surface is obtained immediately [10,11].

If the fractal dimension of a surface is determined by calculating the structure function of two-dimensional profiles the fractal dimension is obtained by

$$D_{OF} = D_{Profile} + 1 \qquad (2.4)$$

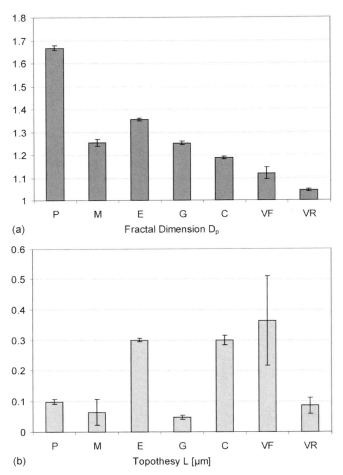

(a) Fractal Dimension D_p

(b) Topothesy L [µm]

Figure 2.3 Fractal dimension D_p (a) and topothesy L (b) for various surface modifications, computed using the structure function $S(\sigma)$ calculated from the tracing profiles (Hommel tester, 1500 mm tracing length, 0.5 mm s^{-1} scan rate): P, polished; M, machined; E, etched; G, glass bead blasted; C, corundum blasted; VF, vacuum plasma sprayed fine; VR, vacuum plasma sprayed rough.

This simplification can only be used if anisotropic surfaces are investigated, i.e. the structure function must be independent of the direction the tracing profiles were measured.

In Figure 2.3 results for the fractal dimension D_f and topothesy L for the same surface modifications as in Figure 2.1 (polishing, machining, irradiation with various materials, etching and plasma spraying) are shown.

Unfiltered profile data were used for this computation so that all possible profile information could be evaluated. The results obtained from this computation did not meet our expectations in that the highest values obtained were for the fractal dimension of the polished surfaces, whereas the lowest values were for the highly structured vacuum plasma-sprayed surfaces. These results could be attributable to the fact that the tracing geometry on which the tracing method is based, and the limited horizontal resolution, distorts the reproduction of the true surface structure to some extent. If this is the case, then small details, sharp edges, recesses, overhangs, porosities and narrow channels would either be omitted or smoothed out.

It proved difficult to evaluate the topothesy parameter since the literature contains little information on its physical significance. Inasmuch as this parameter establishes a correlation between vertical roughness and horizontal resolution, it can be regarded as a lateral roughness parameter [12].

2.2
Optical Characterization

Ulrich Beck and Regina Lange

Evaluation of Transverse Microsections

As mentioned in the previous section, tracing profiles are of only limited use for the characterization of structured surfaces, owing to the fact that they can in some cases yield skewed data. Methods such as SEM imaging of transverse microsections provide higher quality images (Figure 2.4). However, in such cases arithmetic evaluation based on conventional tracing profile algorithms is unrealizable.

Specific algorithms can be used to compute the fractal dimension $D_{F,B}$ on the basis of the boundary lines of transverse microsections. The box counting method was used in the present case (Kolmogorov dimension [13]). For that purpose, SEM images of transverse sections were processed digitally in such a way as to extract boundary lines from the images. Then the image resolution was reduced incrementally and for every resolution selected the number of the screen dots needed to visualize the boundary lines were counted [14,15]. The power law

$$N(r) = \text{const} \times r^{-D_{F,B}} \tag{2.5}$$

between the edge length of screen dot r and the number of screen dots $N(r)$ needed to visualize the boundary line applies to the boundary line fractality.

This power law can be traced back to a linear relationship via logarithmic calculations. In that way the slope of the regression line in a double logarithmic

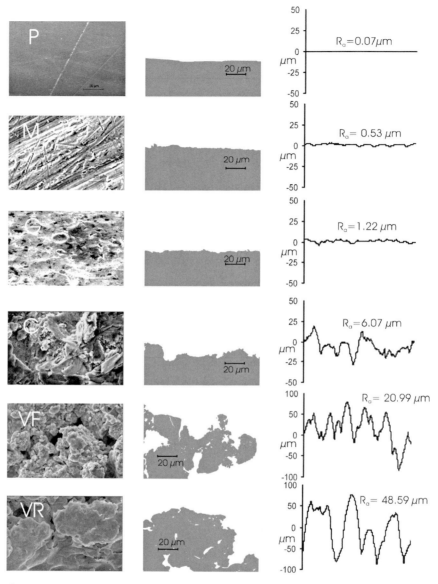

Figure 2.4 Comparison of SEM imagery (S360 Cambridge), transverse microsections (FEI Quantax) and tracing profiles (Hommel tester, 1500 mm tracing length, 0.5 mm s^{-1} scan rate) for various surface modifications: P, polished; M, machined; G, glass bead blasted; C, corundum blasted; VF, vacuum plasma sprayed fine; VR, vacuum plasma sprayed rough.

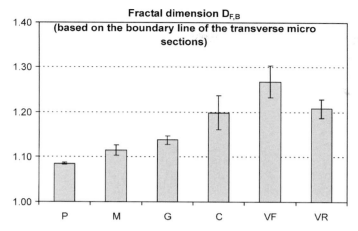

Figure 2.5 Fractal dimension values $D_{F,B}$ for various surface modifications (based on the boundary line of the transverse micro sections): P, polished; M, machined; G, glass bead blasted; C, corundum blasted; VF, vacuum plasma sprayed fine; VR, vacuum plasma sprayed rough.

graph directly indicates the value for fractal dimension $D_{F,B}$ of the boundary line. As in calculation of the fractal dimension with the help of the structure function the fractal dimension of the corresponding surface is obtained by Eq. (2.4) (only for anisotropic surfaces).

Figure 2.5 shows results for the fractal dimension $D_{F,B}$ obtained with the above described method for a series of surface modifications. In contrast to tracing profile values, the anticipated positive correlation between increased fractal dimension values and greater roughness is observed. However, inasmuch as the values lie extremely close to one another, they do not adequately reflect the structure character of the individual surfaces.

Evaluation of SEM Stereo Image Sets

The evaluation of SEM stereo image sets is a far more flexible and precise method than the two methods described above. In this method, a set of stereo images is generated via eucentric tipping of the sample table in the SEM instrument, which can be used to compute a three-dimensional model of the sample surface [16]. Figure 2.6 shows an example for a calculated digital surface model of a machined surface. Owing to the SEM image's high resolution and outstanding depth of focus, this model provides far more accurate results than conventional tracing devices, which also means that the parameters computed from these results will be more differentiated and will reflect reality more accurately.

In addition to a highly precise eucentric SEM tipping apparatus, professional software (Alicona-MeX) is used for the extensive computation required by the evaluation process. However, none of our own results are as yet available as we are currently in the process of defining the method.

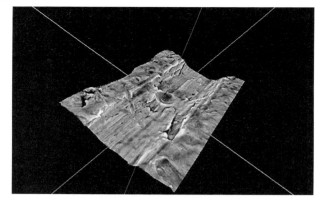

Figure 2.6 Digital surface model of a machined titanium surface calculated from a SEM stereo pair with the help of Alicona MeX software.

Discussion

Tracing-based topography characterization of structured surfaces is a commonly used method that is nondestructive and relatively flexible, and whose profile data can be used to compute a broad range of parameters. However, the quality and quantity of information provided by this parameter is determined by the quality of the profile data, which is in turn strongly influenced by the measurement parameters selected (measurement section, measuring velocity, horizontal and vertical resolution, filter cutoff wavelength, measurement location and direction in the case of anisotropic surfaces) and the surface itself (e.g. porosities).

Furthermore the method is not suitable for very pressure-sensitive surfaces since the surface must be touched in order to measure it. Such surfaces should be measured using optical methods such as laser interferometry.

Compared to the tracing method, where a great number of profiles for the surface area under investigation can be recorded and evaluated without any difficulty for purposes of identifying local differences, the making and investigation of transverse sections is extremely labor intensive. Here often the section orientation can have a major impact on subsequent results, particularly for highly structured or isotropic surfaces. The SEM image enlargement size and resolution can also affect the results. However, as this is a destructive and extremely labor-intensive procedure, it has little practical relevance.

A good alternative option to get precise results for topographic surface data can be the evaluation of SEM stereo image pairs.

Of the great number of surface parameters that can be calculated from surface height data, the fractal dimension is a very interesting parameter because technical surfaces often have self-similar properties. Some fractal parameter computation methods from surface height data were already presented in this section. There are also some other methods for determining fractal dimension, for example by electrochemical methods (see Section 2.3). However, these fractal parameters can only form the basis for comparisons of differing surface structures if the same analytic method is used for all target surfaces. Neither the fractal dimension nor the topothesy parameter provides a clear structure characterization of random fractals (i.e. natural surfaces). Hence in order to fully describe surface structure, it is necessary to combine a maximum number of structure-characterizing parameters with each other.

2.3
Electrochemical Characterization

2.3.1
DC Methods

Ulrich Beck, Regina Lange, and Roger Thull

Electrochemical methods appear to be particularly suitable for characterization of the interface between a material and a biosystem, by virtue of the fact that contact with a suitable electrolyte such as phosphate buffered saline (PBS) during an *in vitro* experiment can accurately simulate real conditions in the human organism. Consequently, the physiological saline solution (pH = 7.2) was used as an electrolyte for the most of the electrochemical experiments described in the present report.

Electrochemical methods characterize the dynamics of the electrical and electrochemical processes that occur at the interface between a biomaterial and a biosystem, and allow for (a) visualization of these processing using a fast-motion format by simply varying the electrical field; or (b) simulation of unusual stress states. Although all of the methods described in the present report are integral in nature, they can also be applied to specific spatial resolutions through the use of microprobes.

All of our electrochemical investigations were realized by applying a three-electrode technique to various titanium surface modifications (see Chapter 3 of Part I). The titanium sample under investigation was integrated as working electrode; a 4.67 cm^2 platinum plate at a distance of approximately 30 mm from the working electrode was used as counter electrode (auxiliary electrode); and a saturated calomel electrode (SCE) served as a reference system. The measurements were realized using an ECO-Chemie Autolab system and a Zahner-Elektrik IM6E [17].

Measurement of Open Circuit Potential
Modifying an electrode's anodic or cathodic response (e.g. by changing the chemical composition of the biomaterial or electrolyte) results in an open circuit potential (E_{OCP}) shift. During each experiment, this mechanism was observed and documented in the PBS electrolyte for 30 min.

Stationary Current–Voltage Characteristic (Linear Voltammetry);
Corrosion Parameter Computation
Current (and calibrated current density)–potential curves reflect either substance conversion at the phase boundary due to the applied electrical field, or electrical potential shift induced by mass transfer. This in turn allows for precise detection of electrochemical passivity (passive areas), as well as highly corroded areas on the electrode (transpassive areas, pitting, etc.).

Exchange current density (and corrosion current density, which means for example the extent of electrode corrosion that occurs without any external applied field) at the point of open circuit potential (equilibrium rest potential) can be determined using the TAFEL method. The slope of the current–voltage curve at equilibrium rest potential provides information on corrosion resistance. Using measurement loops, a determination can be made as to whether and when electrodes after a corrosive breakdown become passive again in the anodic area when current flows back in the cathodic direction.

Assuming that both the anodic and cathodic reactions remain unchanged, the exchange current density and corrosion resistance values can be used to measure increases in both real and electrochemically active surfaces. However, inasmuch as it cannot always be ruled out that changes in the anodic and cathodic reactions may also occur and promote altered exchange current density or corrosion resistance, it is necessary to realize accompanying investigations such as SEM image capture, chronoamperometric measurements or interface capacity measurements.

The current–voltage characteristics were recorded in a potential range of approximately $E_{ocp} - 250\,\text{mV} \leq E \leq E_{ocp} + 250\,\text{mV}$ with a scan rate of $0.5\,\text{mV s}^{-1}$. The geometric electrode surface was kept constant ($A = 2.27\,\text{cm}^2$) for all surface modifications. Thus all modifications could be compared among each other without referencing the measured current to geometric area (calculation of current density). In Figure 2.7 is shown a comparison between corrosion current and corrosion resistance for various surface modifications (according to Chapter 3 of Part I).

Chronoamperometry

Another method for measuring the increase in an electrochemically active surface is chronoamperometry. In this method, the change in the electrical displacement flux and surface charge induced by a potential jump of 300 mV in anodic direction was measured through integration of the current–time characteristic. The resulting value ΔQ can also be used to measure surface area increasing and correlates well with the increase in exchange current density (Figure 2.8).

Electrochemical Measurement of the Fractal Dimension

Use of the fractal dimension parameter is not limited to mere topographical characterization of rough surfaces: the fractal character of rough surfaces can also shed greater light on numerous electrochemical reaction mechanisms. Moreover, diffusion on rough surfaces is also governed by fractal laws [18–20]. The following generalized Cottrell equation is obtained by incorporating the fractal dimension parameter D_f into the Cottrell equation, which describes the time dependency of diffusion controlled current $i(t)$:

$$i(t) = \sigma_F t^{-a} \quad \text{with } a = (D_f - 1)/2 \text{ and } \sigma_F \text{ is the fractal Cottrell coefficient}$$

(2.6)

This equation also covers the special case for a smooth surface where

$$D_f = 2(a = 1/2)$$

(2.7)

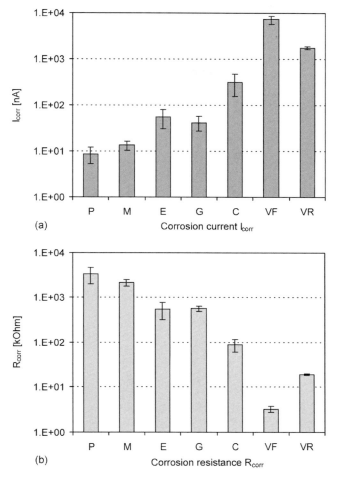

(a)

(b)

Figure 2.7 Results of linear voltammetry (measured in PBS (pH = 7,2), $E_{ocp} - 250\,\text{mV} \leq E \leq E_{ocp} + 250\,\text{mV}$ (vs. SCE), scan rate of $0.5\,\text{mV s}^{-1}$, $A = 2.27\,\text{cm}^2$). Corrosion current I_{corr} (a) and corrosion resistance R_{corr} (b) for various surface modifications: P, polished; M, machined; E, etched; G, glass bead blasted; C, corundum blasted; VF, vacuum plasma sprayed fine; VR, vacuum plasma sprayed rough.

Application of this law to cyclic voltammetry (Riemann Liouville transformation) results in the following relationship [21]:

$$i_{peak} \propto v^a \tag{2.8}$$

This means that if the reaction occurs in a diffusion controlled manner, a potential relationship exists between the maximum current value i_{peak} in the cathodic branch of the voltagram on the one hand and the scan velocity $v = \mathrm{d}E/\mathrm{d}t$ and the parameter α

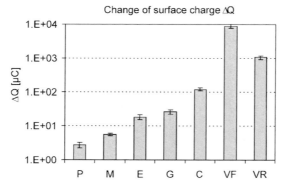

Figure 2.8 Results of chronoamperometry (measured in PBS
(pH = 7.2), potential jump $E_{ocp} \rightarrow E_{ocp} + 300$ mV (vs. SCE),
$A = 2.27$ cm²). Change of surface charge ΔQ for various surface
modifications: P, polished; M, machined; E, etched; G, glass bead
blasted; C, corundum blasted; VF, vacuum plasma sprayed fine;
VR, vacuum plasma sprayed rough.

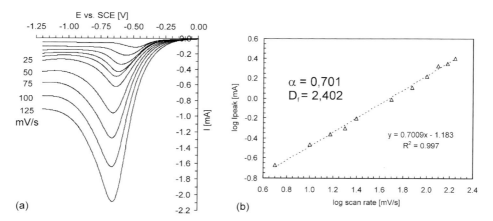

Figure 2.9 Electrochemical determination of fractal dimension
(linear voltammetry with variation of scan rate in 0.5 m
Na₂SO₄ + 10 mM K₃[Fe(CN)₆], $A = 2.27$ cm²). (a) Correlation
between maximum current i_{peak} and scan rate for a corundum-
blasted Ti surface. (b) Determination of the fractal dimension
based on the double logarithmic curve $\log(i_{peak}) = f(\log(\nu))$.

on the other hand (and thus to the fractal dimension D_f). This allows for determination of the parameter α as the slope of the regression line (and thus also allows for
determination of the fractal dimension D_f) by varying the scan rate and subsequent
double logarithmic plotting of $\log i_{peak} = f(\log(\nu))$ (Figure 2.9).

The measurements were realized in 0.5 M Na₂SO₄ using the redox couple $Fe^{2+/3+}$
(10 mM K₃[Fe(CN)₆]. The scanning rate was varied from 5 to 200 mV s⁻¹ and

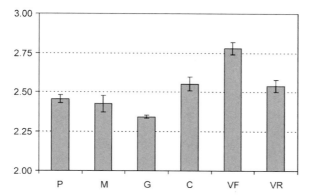

Figure 2.10 Results of electrochemical determining of fractal dimension D_f (linear voltammetry with variation of scan rate in 0.5 m Na_2SO_4 + 10 mM $K_3[Fe(CN)_6]$, $A = 2.27\,cm^2$) for various surface modifications: P, polished; M, machined; E, etched; G, glass bead blasted; C, corundum blasted; VF, vacuum plasma sprayed fine; VR, vacuum plasma sprayed rough.

scanning was realized in the cathodic direction in a potential range of 0.2 to $-1.2\,V$ (vs. SCE). The results obtained with this method for various rough titanium surfaces are shown in Figure 2.10 [22].

The polished surface value is surprisingly high. A tendency is also visible similar to that observed for the values measured using transverse sections, except that the electrochemical method is easier to realize and is nondestructive. Furthermore, a value computed on the basis of the entire measured surface is already obtained here, and thus the measurement location has only a negligible impact on the measurement results compared to the effect of the transverse sections or the tracing procedure. All surface irregularities were documented via electrolyte wetting. The high value obtained for the polished surface may be attributable to the dominant influence of the passive coating.

2.3.2
Impedance Spectroscopic Interface Characterization: Titanium in Contact with Electrolyte and Biosystem

Ulrich Beck, Regina Lange, and Roger Thull

In addition to the DC methods described above, AC methods such as electrochemical impedance spectroscopy (EIS) can be effective for characterizing the interface between a material and a biosystem. An AC impedance frequency spectrum $Z = f$ (ω), where $\omega = 2\pi f$, can be recorded by applying of a variable frequency, low-amplitude and sinusoidal potential change to the phase boundary, and by measuring the resulting current response, which will reflect the dynamics of the interface. As is done in electrical engineering, the individual elements (capacities and resistances) can be joined together into equivalent circuits of varying complexity to model the interface dynamics, which is primarily determined by capacities (oxide coating and electrochemical double coating) and resistances (electrolyte resistance, throughput and polarization resistance). In addition to the classic electrical engineering components, other elements are available that describe specific electrochemical reactions and processes such as the Warburg element (diffusion-controlled reaction), as well as the constant phase element (CPE), which describes frequency dispersion.

In order to find at a suitable equivalent circuit that is also coherent from a physicochemical standpoint and that can be used to model the measured frequency spectrum, it is necessary to obtain information beforehand regarding the system that is to be modeled since frequency spectrums tend to be ambiguous and modeling can be realized on the basis of more than one equivalent circuit, although this not always makes sense from a physical standpoint. Thus under some circumstances supplementary information regarding the entity under investigation must be obtained via accompanying methods (e.g. SEM, transverse sections or electrochemical DC methods) and incorporated into the modeling process using an equivalent circuit.

Application of Impedance Spectroscopy to Rough Surfaces

The use of impedance spectroscopy on rough titanium surfaces can give rise to problems. Titanium oxidizes spontaneously with ambient oxygen in such a way that there is always a thin passive coating on the surface of the metal. More importantly, this is a property that makes titanium a particularly suitable biomaterial, although for purposes of the present report the property tends to be a drawback since the oxide coating is also always included in any measurements that are realized. Furthermore the electrochemical double layer always exists if the sample surface is in contact with an electrolyte. So two capacities exist that cannot always be differentiated in the frequency spectrum as two separate time constants according to oxide thickness and properties. For investigations of a *smooth* test sample with a thin passive coating, a series connection of the passive and electrochemical double layer can be assumed, although the substantially smaller capacity of the double coating (*ca* $15\,\mu F\,cm^{-2}$) governs overall surface dynamics and is identifiable as individual time constants in the phase spectrum. In this case, modeling can be realized using a very simple equivalent circuit via parallel connection of the polarization resistance and double-coating capacity. However, when a *rough* surface is being investigated, the following effects are observed (at first glance) in the frequency spectrum that do not occur in smooth surfaces:

1. Surface enlargement increases the electrochemical capacity of the double layer $C \propto A$.
2. At the same time capacity frequency dispersion is observed, i.e. capacity correlates with frequency, and the rougher and more inhomogeneous the surface, the stronger this correlation becomes. This mechanism can be described using a model that integrates a constant phase element, whereby the latter's exponent comprises a unit of measurement for frequency dependency and thus for surface

roughness [23]. There are also variants on this approach in which the exponent of the constant phase element is pegged very closely to the fractal dimension of self-similar surfaces [24–27].

3. Polarization resistance declines with increasing roughness due to surface increasing.

Frequency dispersion precludes any immediate calculation of surface expansion as per item (1) above since the various frequency dependencies have to be considered. It is common practice in such cases to normalize CPE capacities at a calibration frequency of 1 kHz. However, this value of 1 kHz was selected more or less at random, and thus it is possible that the proportions between capacities will be different at other frequencies and that this gap will grow wider as roughness (and with it frequency dispersion) increase.

Another problem to which the interpretation of the spectra of rough titanium surfaces can give rise is of course that surface roughening can also alter the passive coating with the result that, unlike with smooth surfaces, the presence of a compact and reasonably homogeneous oxide coating can no longer be assumed. This in turn can alter the proportions between the passive and electrochemical double layer in such a way that both capacities affect frequency spectrum properties, e.g. the spectrum may then contain multiple time constants. This in turn increases model (equivalent circuit) complexity since it can no longer be assumed that the two capacities are series-connected. Internal passive coating inhomogeneity also induces frequency dispersion in the oxide coating capacity.

Normally impedance spectra of rough surfaces can be modeled on the basis of one or more constant phase elements, although the impact of the oxide and electrochemical double coatings cannot always be clearly differentiated in the presence of more extensive and varied frequency dispersion.

Thick surface coatings (anodic oxidation, plasma method) also result in the occurrence of an additional time constant in the impedance spectrum.

Results

The impedance spectroscopy investigations were realized in PBS electrolyte for a broad spectrum of rough titanium surfaces (polished, machined, glass bead blasted, corundum blasted, microplasma treated, and vacuum plasma sprayed; see Chapter 3 of Part I) at a constant operating point potential (equilibrium rest potential E_{ocp}). The potential was modulated with a sinusoidal potential change in a frequency range from 1 MHz to 10 kHz with an amplitude of 10 mV (single sinus). The same geometric test sample area (2.27 cm^2) was used for all tests. The frequency spectrum readings are shown in Figures 2.11 and 2.13).

Figure 2.11 clearly shows that the rougher the surface, the higher its capacity and the lower its polarization resistance. The spectra for relatively smooth surfaces (i.e. polished, machined and glass bead blasted) can be modeled using an extremely simple equivalent circuit (Figure 2.12a). A qualitative spectral change was observed in the corundum-blasted and plasma-sprayed test samples relative to their "smooth" counterparts (the occurrence of multiple time constants), which in turn produced a

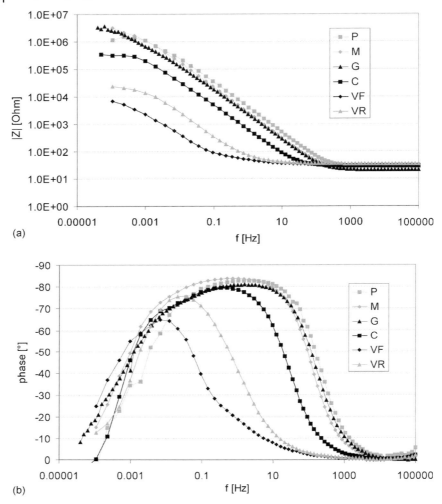

Figure 2.11 Impedance spectra for various surface modifications:
(a) modulus; (b) phase shift (EIS in PBS (pH = 7.2), $E = E_{ocp}$,
$\Delta E = 10$ mV, $f = 1$ MHz to 10 kHz (single sinus)): P, polished;
M, machined; E, etched; G, glass bead blasted; C, corundum
blasted; VF, vacuum plasma sprayed fine; VR, vacuum plasma
sprayed rough.

more complex equivalent circuit (Figure 2.12b and c). Although the occurrence of
multiple time constants in the corundum-blasted surfaces appears to be attributable
to the increased frequency dispersion occasioned by extremely rough surfaces, in
the plasma sprayed surfaces these constants result from the formation of a surface
coating. In the equivalent circuit, this translates into the series connection of two
RC elements [28].

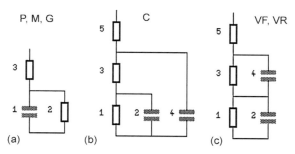

Figure 2.12 Equivalent circuits for the following test sample surfaces: (a) relatively smooth (polished, machined, glass bead blasted); (b) corundum blasted; (c) vacuum plasma sprayed (fine and rough).

For example, for a microplasma-treated titanium test sample the occurrence of multiple time constants resulting from the presence of a surface coating is clearly observable. The fact that two local maximums occurred in the phase spectrum (Figure 2.13a) points to the presence of a surface coating that is "thicker" than a natural passive titanium coating. This supposition was substantiated through realization of a transverse section (Figure 2.13b). After removal of this thick surface coating the second time constant disappeared and the spectrum reverted to that of a "smooth" test sample [29].

The increase in the electrochemically active surface relative to a polished surface can be measured using the capacity obtained via impedance spectroscopy. However, in the case of complex equivalent circuits (i.e. with two or more capacities) and more extensive capacity frequency dispersion, the results have to be verified through comparison with other surface measurement methods (corrosion current measurement, chronoamperometry). Figure 2.14 shows a comparison of the three surface measurement methods for rough titanium surfaces. All of these methods provide relatively consistent results.

In contrast to capacity-based surface enlargement measurement providing relatively accurate results, it is more difficult to compare the CPE exponents for various rough surfaces if their frequency spectra cannot be described using the same equivalent circuit and if no clear assignment of physical and electrochemical phenomena to the individual equivalent circuit elements can be found. However, in the titanium surfaces investigated here, the CPE exponents tended to decrease as roughness increased. CPE coefficient-based computations of the fractal dimension according to [26] and comparing the results of these computations with other fractal dimension measurement methods (digital analysis of boundary lines (see Section 2.2) and the electrochemical method (see Section 2.3.1)) showed similar tendencies for the surface modifications investigated, although some of the individual absolute values differed considerably (Figure 2.15). Particularly noteworthy here were the relatively high electrochemical LSV measurement values, whereas the image processing and EIS values were comparably.

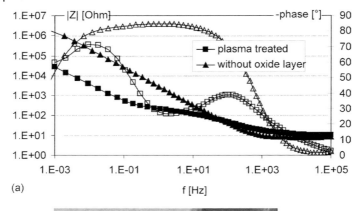

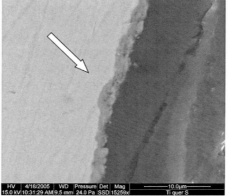

(a)

(b)

Figure 2.13 (a) Impedance spectrum of a microplasma-sprayed titanium sample (with and after removal of the surface layer). (b) SEM image of a transverse section of the microplasma-treated sample with a 1–2 μm surface coating.

Discussion

Electrochemical methods are highly suitable for the characterization of metal biomaterials since they are (a) nondestructive; and (b) accurately simulate the relationships in the human body, providing that a suitable electrolyte is selected. Thus the present tests provided good surface enlargement measurement results for rough titanium surfaces. A relatively new method for determining of a fractal parameter based on electrochemical methods also yielded good results.

The results provided by electrochemical impedance spectroscopy can in some cases be ambiguous, particularly when relatively complex surfaces (very rough surfaces, "thick" oxide coatings, surface coatings) are involved. In such cases, the measurement results should be assessed critically and possibly validated through the use of accompanying measurement methods.

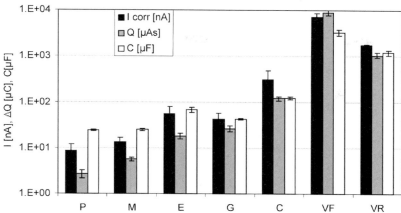

Figure 2.14 Comparison of the results of the various surface enlargement measurement methods (linear voltammetry (I_{corr}), chronoamperometry (ΔQ) and electrochemical impedance spectroscopy (C)) for various surface modifications: P, polished; M, machined; E, etched; G, glass bead blasted; C, corundum blasted; VF, vacuum plasma sprayed fine; VR, vacuum plasma sprayed rough.

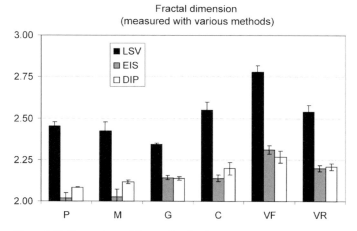

Figure 2.15 Comparison of the results of various measurement methods for the fractal dimension D_f (linear voltammetry with variation of scan rate (LSV), electrochemical impedance spectroscopy (EIS) and digital image processing of the boundary line (DIP)) for various surface modifications: P, polished; M, machined; E, etched; G, glass bead blasted; C, corundum blasted; VF, vacuum plasma sprayed fine; VR, vacuum plasma sprayed rough.

References

1 Bigerelle, M., Anselme, K., Dufresne, E., Hardouin, P. and Iost, A. (2002) An unscaled parameter to measure the order of surfaces: a new surface elaboration to increase cells adhesion. *Biomol. Eng.*, **19**, 79–83.

2 Hasegawa, M., Liu, J., Okuda, K. and Nunobiki, M. (1996) Calculation of the fractal dimensions of machined surface profiles. *Wear*, **192**, 40–45.

3 Russ, J.C. (1998) Fractal dimension measurement of engineering surfaces. *Int. J. Mach. Tools Manufact.*, **38**, 567–571.

4 Chauvy, P.-F., Madore, C. and Landolt, D. (1998) Variable length scale analysis of surface topography: characterization of titanium surfaces for biomedical applications. *Surf. Coat. Technol.*, **110**, 48–56.

5 Russ, J.C. (1994) The relationship between boundary lines and surfaces, *Fractal Surfaces*, Plenum Press, New York, pp. 59.

6 Berry, M.V. (1979) Diffractals. *J. Phys. A: Math. Gen.*, **12**, 781–797.

7 Bates, J.B. and Chu, Y.T. (1988) Surface topography and electrical response of metal–electrolyte interfaces. *Solid State Ionics*, **28–30**, 1388–1395.

8 Bates, J.B., Chu, Y.T. and Stribling, W.T. (1988) Surface topography and impedance of metal–electrolyte interfaces. *Phys. Rev. Lett.*, **60**, 627–630.

9 Russ, J.C. (1994) Hurst and Fourier analysis, in *Fractal Surfaces*, Plenum Press, New York, pp. 83.

10 Amini, N., Rosén, B.-G. and Thomas, T.R. (1998) In: *Fractals and beyond* (Ed.: M.M. Novak), World Scientific Publ., Singapore, pp. 183.

11 Schmähling, J., Hamprecht, F.A. and Hoffmann, D.M.P. (2006) A three-dimensional measure of surface roughness based on mathematical morphology. *Int. J. Mach. Tools Manufact.*, **46**, 1764–1769.

12 Jahn, R. and Truckenbrodt, H. (2004) A simple fractal analysis method of the surface roughness. *J. Mat. Proc. Technol.*, **145**, 40–45.

13 Russ, J.C. (1994) Mosaic amalgamation and the Kolmogorov dimension, in *Fractal Surfaces*, Plenum Press, New York, pp. 38.

14 Kirbs, A., Lange, R., Nebe, B., Rychly, J., Müller, P. and Beck, U. (2003) On the description of the fractal nature of microstructured surfaces of titanium implants. *Mater. Sci. Eng. C*, **23**, 413–418.

15 Amada, S. and Hirose, T. (1998) Influence of grit blasting pre-treatment on the adhesion strength of plasma sprayed coatings: fractal analysis of roughness. *Surf. Coat. Technol.*, **102**, 132–137.

16 Ritter, M., Sinram, O., Albertz, J. and Hohenberg, H. (2003) Quantitative 3D reconstruction of biological surfaces. *Microsc. Microanal.*, **9**, 476–477.

17 Kirbs, A., Lange, R., Nebe, B., Rychly, J., Baumann, A., Neumann, H.-G. and Beck, U. (2003) Methods for the physical and chemical characterization of surfaces of titanium implants. *Mater. Sci. Eng. C*, **23**, 425–429.

18 Nyikos, L. and Pajkossy, T. (1986) Diffusion to fractal surfaces: short communication. *Electrochim. Acta*, **31**, 1347–1350.

19 Pajkossy, T. and Nyikos, L. (1989) Diffusion to fractal surfaces: II. Verification of theory. *Electrochim. Acta*, **34**, 171–179.

20 Pajkossy, T. and Nyikos, L. (1989) Diffusion to fractal surfaces: III. Linear sweep and cyclic voltammograms. *Electrochim. Acta*, **34**, 181–186.

21 Strömme, M., Niklasson, G.A. and Granqvist, C.G. (1995) Voltammetry on

fractals. *Solid State Commun.*, **96**, 151–154.

22 Kirbs, A., Lange, R., Nebe, B., Rychly, J. Müller, P. and Beck, U. (2003) On the description of the fractal nature of microstructured surfaces of titanium implants, *Mater. Sci. Eng. C*, **23**, 413–418.

23 de Levie, R. (1965) The influence of surface roughness of solid electrodes on electrochemical measurements. *Electrochim. Acta*, **10**, 113–130.

24 de Levie, R. (1989) On the impedance of electrodes with rough interfaces. *J. Electroanal. Chem.*, **261**, 1–9.

25 Liu, S.H. (1985) Fractal model for the ac response of a rough interface. *Phys. Rev. Lett.*, **55**, 529–532.

26 Pajkossy, T. and Nyikos, L. (1986) Impedance of fractal blocking electrodes. *J. Electrochem. Soc.*, **133**, 2061–2064.

27 Rammelt, U. and Reinhard, G. (1990) On the applicability of a constant phase element to the estimation of roughness of solid metal electrodes. *Electrochim. Acta*, **35**, 1045–1049.

28 Kirbs, A., Lange, R., Nebe, B., Rychly, J. Baumann, H.-G. and Beck, U. (2003) Methods for the physical and chemical characterization of surfaces of titanium implants, *Mater. Sci. Eng. C*, **23**, 425–429.

29 Beck, U., Lange, R. and Neumann, H.-G. (2007) Micro-plasma textured Ti-implant surfaces. *Biomol. Eng.* **24**, 47–51.

3
Spectroscopic Interface Characterization

3.1
Electronic Structure of the Interface

Andreas Zoll and Roger Thull

Materials and Methods

Titanium oxide films of different thickness were prepared using a custom made radio frequency (r.f.) magnetron sputtering system. The background pressure was 2×10^{-5} mbar obtained using a turbomolecular pump backed by a mechanical pump. High-purity argon and oxygen were used as sputtering and reactive gases. The target was either a titanium disc (purity 99.9%) of 12.95 cm diameter or a titanium oxide disc (purity 99.998%) of the same diameter. The system was powered by an r.f. generator at a frequency of 13.56 MHz. The discharge was generated at a constant r.f. power of 300 W and the flow rates of Ar and O_2 were kept at constant values of 100 and 20 sccm respectively when using the titanium target. If the TiO_2 target was used the flow rate was 180 sccm Ar.

The stoichiometry of titanium oxide layers deposited on titanium discs (grade 2, 15 mm diameter) was determined by X-ray diffraction (XRD) using grazing incidence because of the small thickness of the TiO_2 films. The distance between samples and target was 60 mm.

Film thickness and with it the rate of growth was measured using atomic force microscopy (AFM). As the electronic structure of the TiO_2 surface should be analyzed in contact with aqueous electrolytes, most methods of modern surface characterization could not be applied, because they require ultrahigh-vacuum conditions. So, electrolytic electroreflectance spectroscopy (ERS) was used to study the electronic structure of the interface *in situ*.

The term electroreflectance is understood to mean the change in the reflectance of a solid body caused by the effects of an electric field. More precisely an electric field lifts the translation symmetry along its direction and tilts the bands spatially. This appears as a change of the optical constants (and with it the dielectric function) in consequence of the Franz–Keldysh effect, which results in an increase or decrease in the reflected light. The recording of the field-dependent change in the reflectance as a

Metallic Biomaterial Interfaces. Edited by J. Breme, C. J. Kirkpatrick, and R. Thull
Copyright © 2008 WILEY-VCH Verlag GmbH & Co. KGaA, Weinheim
ISBN: 978-3-527-31860-5

function of the energy of the incident light delivers the electroreflectance spectrum. To eliminate the influence of the energy-dependent light intensity, the ERS signal is normalized to the intensity of the unmodulated signal.

This technique permits better visualization of structures in optical spectra than in reflectance measurements because ERS corresponds to the third derivative of the primary spectra with respect to E [4].

Because of this derivative-like nature, sharp spectral features can be observed in modulated reflectance spectra of semiconductors. With these spectra it is possible to analyze the properties of the material under study. Changes in reflectance as small as 10^{-6}–10^{-7} can be observed using phase-sensitive techniques. The ability to perform a line shape fit is one of the great advantages of ERS.

Figure 3.1 shows the setup used for ERS. The surface was located in a measuring cell continuously perfused with electrolyte and provided with a steel counter electrode, which was also used as reference electrode. The electrode potential was varied within the range 0 to $+3$ V.

On the DC voltage, a rectangular AC voltage of small amplitude, in the range of $U_{AC} = 0.05$–1.00 V with a frequency f of between 50 and 500 Hz was superimposed.

The surface is illuminated through a quartz window with monochromatic light in the spectral range $\lambda = 200$–800 nm of perpendicular incidence. The reflected light, consisting of $R_0 + \Delta R$, was detected with a photomultiplier. The light source was a

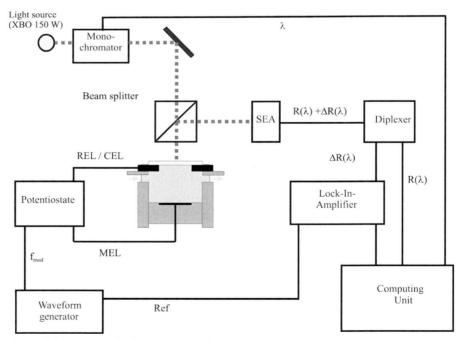

Figure 3.1 Experimental setup for electroreflectance spectroscopy.

150 W xenon high-pressure discharge lamp XBO. The modulated light component ΔR was separated using a frequency separating filter and analyzed with respect to the modulation using a lock-in amplifier.

All data were recorded simultaneously; hence changes in the XBO spectra could not affect ERS measurements.

Characteristic parameters of transitions were extracted from ERS data using the fitting function

$$\frac{\Delta R}{R}(\hbar\omega) = \Re\left[Ce^{i\mu}(\hbar\omega - E_0 + i\Gamma)^{-n}\right] \tag{3.1}$$

where C is the amplitude, θ is a phase angle that accounts for the mixture of real and imaginary components of ε as well as the influence of non-uniform electric fields, interference and electron–hole interaction effects, and E_0 is the transition energy. The parameter Γ describes the broadening and n depends on the type of critical point. As Eq. (3.1) is difficult to handle, the following equation, which is based on Eq. (3.1) was used:

$$\frac{\Delta R}{R}(\hbar\omega) = C[(\hbar\omega - E_0)^2 + \Gamma^2]^{-m/2}\cos(\theta - m\Psi) \tag{3.2}$$

where

$$\Psi = \cos^{-1}\left(\frac{\hbar\omega - E_0}{\sqrt{(\hbar\omega - E_0)^2 + \Gamma^2}}\right) \tag{3.3}$$

The experimental data were fit to Eq. (3.2) using a nonlinear least squares fitting routine.

Results and Discussion

Electroreflectance spectra were studied on TiO_2 coatings between 25 and 100 nm thickness for reactive sputtered films and between 65 and 200 nm in the non-reactive case. XRD of these films showed that reactive sputtered coatings consisted mainly of anatase with a small fraction of rutile. Films prepared with the TiO_2 target consisted of anatase only.

Figure 3.2 shows measured transition energies of reactively sputtered samples. First, none of the samples showed transitions at the intrinsic band gap of TiO_2 at 3.20 eV. Only a small amount of samples showed transitions between 3.0 and 3.2 eV, which is the region of the intrinsic band gap of titanium dioxide bulk material.

Instead of this, most electroreflectance spectra showed sub-band-gap transitions. Thereby for thinner coatings only sub-band-gap transitions could be measured or transitions at the intrinsic band gap. Thicker films, however, showed the intrinsic band gap and a transition at lower energy.

These sub-band-gap transitions were a consequence of a high density of surface states, which resulted from interactions between the surface and the electrolyte. A connection with a reduced density of oxygen atoms in the TiO_2 coating is indeed possible, but would only explain transitions at 2.11 eV [4].

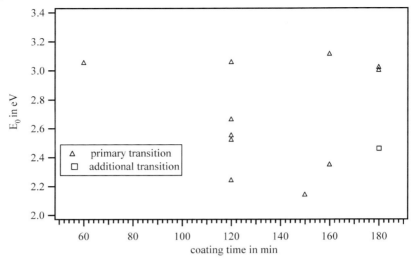

Figure 3.2 Transition energies for reactively sputtered TiO$_2$ layers (parameters for each measurement: $U_{DC} = +2.0$ V, $U_{AC} = \pm0.5$ V, $f = 73$ Hz, $T = 310$ K, electrolyte: PBS).

This is reinforced by measurements at the non-reactive sputtered coatings, which also showed a broad distribution of transition energies, but none at 2.11 eV. Transitions at 2.4 and 2.7 eV correspond to the optical ionization energy of a TiO$_2$ single crystal at 2.33 and 2.75 eV [4].

Non-reactively sputtered TiO$_2$ films showed a similar behavior (Figure 3.3), whereas more samples than in the reactively sputtered case did not show any transitions at all. This may originate from the higher deposition rate with non-reactive deposition, which can be realized either by a higher particle current or by larger particles than in the reactive case. In doing so both cases have similar results: while with a higher particle current relaxation processes at the surface are eliminated by the fast impact rate, it is the particle size that makes diffusion processes impossible at the surface with larger particles. This effect is supported by the short coating time of less than 30 min, which does not cause heating of the sample and does not support crystal growth processes. This is also corroborated by XRD measurements where all non-reactively sputtered samples with a coating time of 30 min showed only peaks from the titanium substrate but none from the TiO$_2$ film.

As in amorphous solid bodies electronic band structures (like in crystalline semiconductors) do not exist, determination of well-defined transition energies is impossible.

All measurements in these two series were performed using the same DC bias and modulation voltages for all samples. So it is interesting that, for example, all samples which were sputtered non-reactively for 15 min, showed different transition energies although they were prepared simultaneously during the whole process. Hence most transitions seem to have their origin in a variety of surface states at the interface between the semiconductive coating and the electrolyte.

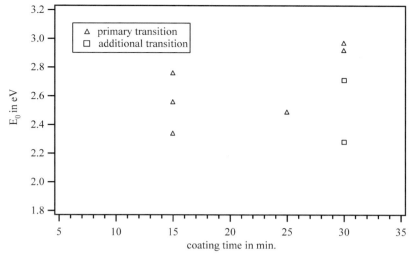

Figure 3.3 Transition energies of TiO$_2$ layers sputtered using the TiO$_2$ target (parameters for each measurement: $U_{DC} = +2.0$ V, $U_{AC} = \pm 0.5$ V, $f = 73$ Hz, $T = 310$ K, electrolyte: PBS).

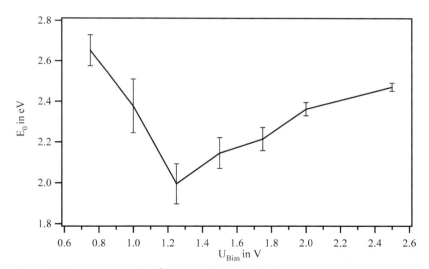

Figure 3.4 Transition energies of one sample measured with different DC bias voltages (parameters for each measurement: $U_{AC} = \pm 0.5$ V, $f = 73$ Hz, $T = 310$ K, electrolyte: PBS).

Another aspect is the dependency of transition energy on the DC bias voltage. Figure 3.4 shows measured transition energies of one sample at different voltages. This dependency is related to a high density of surface states at the semiconductor–electrolyte interface and a consequence of the so called Fermi-level pinning. In this case, variation of bias voltage results in a variation of the energetic position of the Fermi level. Thereby electrons are excited from various, just filled surface states into the conduction band.

If it is assumed that measured energies derive from transitions into the conduction band, a higher bias voltage causes a stronger bending of the conduction band. The outcome of this is a linear increase of transition energy as is shown in the right part of Figure 3.4.

If the density of surface states below the intrinsic band gap is very high, the energetic position of the Fermi level is fixed at the semiconductor surface. Thereby the barrier height between the metal and the semiconductor becomes independent from the metal's work function and is strongly influenced by properties of the semiconductor surface. In this case the barrier height cannot – as usual – be calculated. So a decreasing transition energy for bias voltages smaller than 1.2 V can also be interpreted as an effect of Fermi-level pinning on properties of the metal–semiconductor interface. This also allows an interpretation of the decreasing transition energies for bias voltages smaller than 1.2 V as a consequence of level pinning.

Another indicator for a pinned Fermi level is the decreasing amplitude of the ERS signal for increasing bias voltage. According to Silberstein's argument an increasing bias voltage causes a decrease of modulation voltage in the space charge layer at the semiconductor–electrolyte interface [5]. If the density of surface states relaxing faster than $1/f$ at the interface is very high, this relaxation can be supposed to be bias dependent because of non-ideal dielectric properties of the polycrystalline film. If this density of states at equilibrium is high enough, the energetic position of the Fermi level will be changed.

Conclusions

Electroreflectance spectroscopic studies on titanium oxide thin-film surfaces in contact with aqueous electrolytes have shown complex electronic properties at the surface. Thereby many effects are connected with a high density of surface states, which results in the so-called Fermi-level pinning at the interface between the surface and the electrolyte.

3.2
Spectro-electrochemistry of the Surface

Andreas Zoll and Roger Thull

Materials and Methods

The experimental setup used for these measurements is described in the previous section. The major difference was that for each measurement series only one sample was used, which was cleaned ultrasonically in double-distilled water after every measurement. The measurement cell and the remaining electrolyte system were also cleaned after every measurement to avoid mixture of the different electrolytes.

All other parameters, which might influence measurements, were kept at constant values for all measurements. For this measurement series only reactively sputtered samples could be used because of the weak adhesion of TiO_2 layers sputtered with the TiO_2 target.

Characteristic parameters were extracted as described in the previous section.

Results and Discussion

Starting with isotonic NaCl solution the influence of global ion concentration was the first aspect to be determined. Figure 3.5 shows the amplitude of the ERS signal for different ion concentrations. Neither transition energy nor phase factor nor the broadening parameter showed significant changes in this series. The decreasing amplitude was a consequence of higher field strength at the surface. As ion concentration increases, the conductivity of the electrolyte also increases. The outcome of this is a lower potential drop between the counter electrode and the outer Helmholtz layer, which leads to higher field strengths at the electrostatic double layer at the surface. But as the sample was biased to be positive, increasing field strength causes a decrease in the number of electrons at the surface and with it a decrease of the signal amplitude.

Another aspect was the influence of adsorbed protein molecules at the TiO_2 surface on the ER spectra. After one sample was measured using phosphate buffered saline (PBS) as electrolyte, the sample was put in protein solution for a defined period of time. Therefore it was rinsed with PBS and measured again using the same electrolyte. The choice of PBS as electrolyte for this measurement series was made in order to keep the pH value at a constant physiological level during measurement on the one hand and simulate ion concentration according to the physiological environment on the other hand.

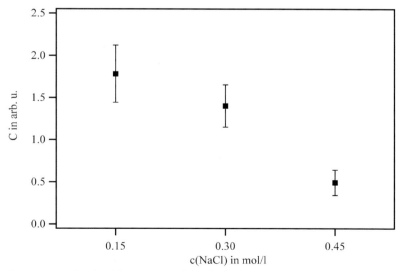

Figure 3.5 Amplitude of the ERS signal for different ion concentrations.

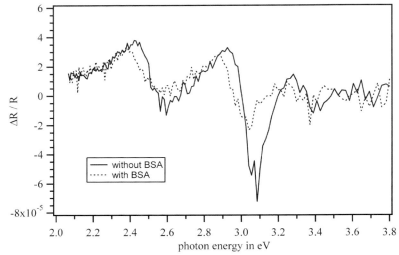

Figure 3.6 Changes in electroreflectance spectra after putting in 1% BSA solution for 30 min at 37 °C (parameters for each measurement: $U_{DC} = +2.0\,V$, $U_{AC} = \pm0.5\,V$, $f = 73\,Hz$, $T = 310\,K$, electrolyte: PBS).

Figure 3.6 shows electroreflectance spectra of a sample with two transitions before and after adsorption of bovine serum albumin (BSA) at the surface. First, a decrease of the amplitude, mainly in the region about 3 eV, is noticeable. This might be a result of an additional potential drop at the protein layer resulting in a lower field strength at the semiconductor surface.

Another aspect is an energetic shift of the transition energies when proteins are adsorbed at the surface. Wrobel et al. indeed report energy shift of transition energies for different bias voltages, but only with transitions below or above the intrinsic energy gap of the semiconductor [6]. Measurement series of the same sample with different bias voltages showed no changes in transition energies, neither below nor above the intrinsic energy gap; hence a pinned Fermi level cannot be considered as possible reason for the energetic shift (see previous section).

For more quantitative studies only samples with one transition were used to determine the influence of adsorbed protein molecules and the time samples were put into protein solutions. Figure 3.7 shows energy shifts of different samples caused by adsorbed protein molecules at the surface; the large error bars have their origin in the relative short-duration measurements. The first distinctive feature is the change in the direction of the energy shift when the protein was changed from BSA to fibrinogen (fraction I from sheep plasma). This effect can eventually be connected with the different surface charge of the protein molecules.

As discussed in the previous section, the amplitude of the ERS signals and – in the case of transition energies below or above the intrinsic energy gap – their energetic position depended on the field strength of the surface. If a positive polarized surface

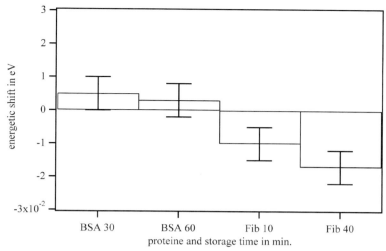

Figure 3.7 Energetic shift of transition energy with bovine serum albumin (BSA) and fibrinogen (Fib) for different periods of time.

gets in contact with an electrolyte, a Helmholtz layer is formed that induces a negative mirror charge in the electrolyte and with it an electric field in the space charge region of the semiconductor. When protein molecules are adsorbed at this surface, the surface charges of these molecules change the electric field.

If protein molecules have negative surface charge like BSA [7], the protein layer causes an increasing field strength inside the semiconductor and with it an energetic shift of the transition energy. If protein molecules have a positive surface charge, the electric field is attenuated, which changes the direction of the energetic shift.

As protein molecules are only adsorbed as a monolayer, a variation of the time samples are put in protein solution should not cause significant changes in energetic shifts, which can also be seen in Figure 3.7.

Conclusion

Effects of changes of electrolyte composition on electroreflectance spectra can be traced back to changes in the field strength in the space charge region of the semiconductor.

References

1 Ellingsen, J.E. (1991) A study on the mechanism of protein adsorption to TiO$_2$. *Biomaterials*, **12**, 593–596.

2 Hanawa, T., Ukai, H. and Murakami, K. (1995) Structure of surface-modified layers of calcium-ion-implanted Ti-6Al-4V and Ti-56Ni. *Mater. Trans.*, **36**, 438–444.

3 Aspnes, D.E. (1972) Direct verification of the third-derivative nature of electroreflectance spectra. *Phys. Rev. Lett.*, **28**, 168.

4 Kim, W.T., Kim, C.D. and Choi, Q.W. (1984) Sub-band-gap photoresponse of TiO$_{2-x}$ thin-film–electrolyte interface. *Phys. Rev. B*, **30**, 3625–3628.

5 Silberstein, R.P., Pollak, F.H., Lyden, J.K. and Tomkiewicz, M. (1981) Optical determination of Fermi-level pinning using electroreflectance. *Phys. Rev. B*, **24**, 7397–7400.

6 Wrobel, J.M. (1986) Effects of dc-bias on electrolyte electroreflectance spectra. *J. Appl. Phys.*, **60**, 368–371.

7 Klinger, A., Steinberg, D., Kohavi, D. and Sela, M. (1997) Mechanism of adsorption of human albumin to titanium *in vitro*. *J. Biomed. Mater. Res.*, **36**, 387–392.

4

Physical Characterization of Protein Adsorption: Time-Resolved Determination of Protein Adsorption Using Quartz Crystal Microgravimetry with Dissipation Monitoring (QCM-D)

Claus Moseke, René Michalek, and Roger Thull

Materials and Methods

The QCM-D experiments were carried out using quartz resonators with different coatings with as-delivered gold-coated crystals (Q-Sense, Sweden) as control samples. Titanium oxide layers were produced by different methods and with different thicknesses in a custom-made magnetron sputtering chamber. Quartz crystals pre-coated with stainless steel (SS2343) were also provided by Q-Sense.

The titanium and titanium oxide coatings were deposited on the quartz resonators by means of magnetron sputtering in a high-vacuum chamber. The deposition of titanium films occurred by acceleration of argon ions, generated in radio frequency plasma, to a disc-shaped target, made from commercially pure (cp) titanium. A special substrate holder was developed in order to coat only the active area of the quartz resonator surfaces. Titanium oxide coatings could be produced by two methods. The oxide could be obtained directly with high deposition rates by sputtering a target disc, made from pressed titanium oxide powder, while the reactive sputtering method used the cp-Ti target and an appropriate reaction gas, consisting of argon for the actual target sputtering and oxygen as reactant gas for the chemical production of the oxide. For better adhesion the oxide layers were not deposited directly on the gold-coated quartz substrate, but superimposed on an intermediate layer of pure titanium. Thus, the higher deposition rate of the direct sputtering method compared with the reactive sputtering was combined with the disadvantage that the former required the opening of the vacuum chamber and the exchange of the target after the deposition of the intermediate titanium layer.

The proteins used for adsorption studies were bovine serum albumin (BSA; from SIGMA), fibrinogen (sheep and human, from SIGMA), and fibronectin (Chemicon). Three antibodies with epitope-specific binding characteristics to fibronectin were applied for conformation studies of already adsorbed fibronectin films. The antibodies were FN-P1H11 (Santa Cruz Biotechnology, specific to the cell binding domain), FN-H300 (Santa Cruz Biotechnology, specific to the C-terminal COO^- region) and FN-MAB (Chemicon, specific to the N-terminal NH_4^+ region).

Metallic Biomaterial Interfaces. Edited by J. Breme, C. J. Kirkpatrick, and R. Thull
Copyright © 2008 WILEY-VCH Verlag GmbH & Co. KGaA, Weinheim
ISBN: 978-3-527-31860-5

All proteins and antibodies were solubilized in phosphate buffer saline at physiological pH, the concentrations being $50\,\mu g\,mL^{-1}$ for fibrinogen and fibronectin and $1\,\mu g\,mL^{-1}$ for the antibody solutions. Cell adsorption experiments were carried out using mouse fibroblasts (cell line L929) cultured in Dulbecco's modified Eagle medium (DMEM) supplemented with 10% fetal calf serum (FCS) at 37 °C in a humidified 5% CO_2 incubator.

Measuring Principle of QCM-D

An AT-cut piezoelectric quartz crystal can be excited to oscillation in the thickness shear mode at a well-defined resonance frequency f by applying a suitable radio frequency voltage across its electrodes [1]. The resonance frequency is determined by the dimensions of the crystal.

A rigid and sufficiently thin layer deposited on the quartz crystal surface causes a relative shift of the resonance frequency, given by

$$\frac{\Delta f}{f} = -\frac{\Delta m}{\rho_Q F d} \tag{4.1}$$

where Δf is the frequency shift, f the resonance frequency, ρ_Q the density of the quartz crystal, F the area of the quartz crystal surface and Δm the mass of the deposited layer [2]. With the introduction of the mass sensitivity C ($17.7\,ng\,cm^{-2}$ Hz^{-1} for a resonator oscillating at 5 MHz) and the number of the overtone n, Eq. (4.1) can be written as

$$\Delta f = -n\frac{1}{C}\Delta m \tag{4.2}$$

the so-called Sauerbrey relation. This formula delivers very accurate results for the thickness determination of rigid and uniform sub-micrometer films produced by physical or chemical vapor deposition processes. The mass sensitivity depends only on the dimensions and the cut direction of the quartz crystal, due to the fact that the additional mass of a sufficiently thin rigid film acts like an increased thickness of the crystal.

This is no longer valid when non-rigid films with viscoelastic properties are deposited in aqueous solution. These films will oscillate elastically with the quartz crystal and dissipate oscillation energy by interaction of molecules inside the film. Furthermore, the incorporation of water molecules from the surrounding solution into the film causes an additional decrease of the resonance frequency, thus leading to the detection of a higher mass than is actually adsorbed [3].

The dampening behavior of a quartz resonator in liquid environment is characterized by the dissipation D, defined as the ratio of energy loss per oscillation and total energy of the oscillating system, which is equal to the conventional quality factor. It is measured concertedly with the frequency by switching the driving power of the oscillator on and off periodically and determining the attenuation constant of the oscillation. A viscoelastic film adsorbed on the resonator surface will increase D by the dissipation shift ΔD, which provides qualitative and – using a Voigt-based

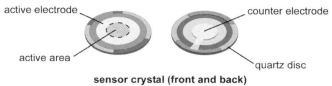

active electrode

counter electrode

active area

quartz disc

sensor crystal (front and back)

Figure 4.1 Standard gold-coated sensor crystal. During the measurement the active front side is in contact with the sample solution, while the back side contains the counter electrode and the electrical contacts. Only adsorption in the active area of the active electrode is sensed in a QCM-D measurement (redrawn after [5]).

non-rigid film model [4] – quantitative information about the physical properties of the film. ΔD depends on density, viscosity and shear elasticity of the film – in the case of multilayer films for every single layer of the film system – and interactions with the surrounding liquid.

The QCM-D setup consists of a measurement chamber containing the sensor crystal (Fig. 4.1), an electronics unit, and a PC with QSoft software. The configuration of the measurement setup for an open flow system is schematically shown in Fig. 4.2. Liquid transport from the sample reservoir to the measurement chamber inlet occurs due to the force of gravity; hence the flow velocity can be easily controlled by adjusting the height of the clamp holding the sample reservoir. The control valve regulates the flow direction of the sample solution, which may either flow from the inlet to the loop outlet via the temperature loop or directly through the sensor cell to the sensor outlet.

Cell adsorption experiments were carried out in a second measurement chamber, featuring autoclavable parts, an observation window and a greater sensor chamber volume that provided more space for the cell culture medium.

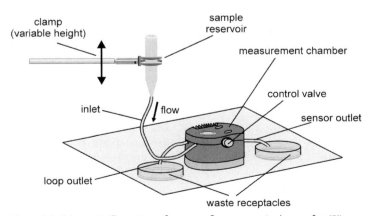

clamp
(variable height)

sample
reservoir

measurement chamber

control valve

inlet

flow

sensor outlet

loop outlet

waste receptacles

Figure 4.2 Schematic illustration of an open flow system (redrawn after [5]).

A typical procedure of measuring starts with the mounting of the sensor crystal in the measurement chamber. After the device has been rinsed with buffer liquid, the f and D values of the sensor crystal in aqueous solution at the given temperature are recorded at the basic resonance frequency f_0 as well as at the overtones $n = 3, 5$ and 7 as a function of time, the stabilized values being set as baselines for the following measurements. Then the sample reservoir is filled with protein stock solution, which then is allowed to flow through the chamber (loop outlet) and – after temperature stabilization – into the sensor chamber, where protein adsorption to the sensor crystal surface takes place. The f and D values are recorded as a function of time during the adsorption process, following rinsing and eventual additions of other adsorptive solutions (e.g. antibody solution).

Protein solutions were prepared in an amount sufficient for all planned experiments and stored at $-20\,°C$ in portions of the size required for one single measurement. Thus, equal concentrations of the sample solutions could be guaranteed for all adsorption experiments. Before a single measurement a portion of the required protein solution was defrosted and tempered to physiological temperature. Quartz crystals – coated and uncoated – were rinsed with isopropanol and blow-dried in nitrogen. The measurement chamber and all attached hoses were rinsed several times with PBS. Emerging air bubbles were carefully removed by pumping the liquid back and forth through the flow system with the help of a syringe.

Results and Discussion

In preliminary experiments BSA showed poor adhesion to the quartz crystal surfaces, indicated by comparatively low frequency shifts Δf. Therefore the subsequent studies were focused on the adsorption of the more adherent proteins fibrinogen and fibronectin.

The aim of the first adsorption studies with sheep fibrinogen was to investigate the influence of protein concentration on the adsorption kinetics. The results of experiments with three different concentrations (25, 500 and 1000 µg mL^{-1}) of fibrinogen in PBS are shown in Fig. 4.3. Apparently, higher protein concentrations caused an increase of the final Δf and ΔD values, suggesting the assumption that the total amount of adsorbed protein increased linearly with protein concentration. However, it must be taken into account that highly concentrated solutions also have higher density and viscosity and therefore underlie stronger interaction with the quartz crystal surface compared to solutions with low protein concentration.

In addition, Figure 4.3 shows different adsorption kinetics for the three protein concentrations during initial adsorption of proteins to the unoccupied gold surface, which took place within the first minute. Derivative analysis showed a correlation of the slope of the adsorption curve with protein concentration. After initial adsorption the slopes decreased to the same value for all protein concentrations.

Fibrinogen of human origin was used for adsorption studies with respect to the different overtones of the basic driving frequency. The Sauerbrey relation (Eq. (4.2)) contains the number n of the overtone as a linear factor; therefore all Δf and ΔD values presented in this study have been normalized to f_0, thereby ensuring comparability of all measurements. Figure 4.4 shows the adsorption curves of human fibrinogen to

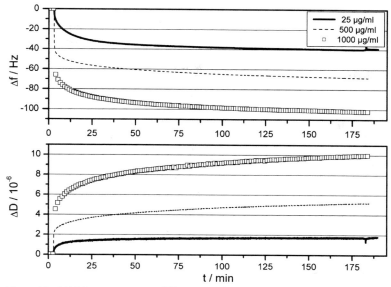

Figure 4.3 QCM-D measurement of fibrinogen adsorption to as-delivered gold-coated quartz crystals from sheep fibrinogen solutions with different concentrations.

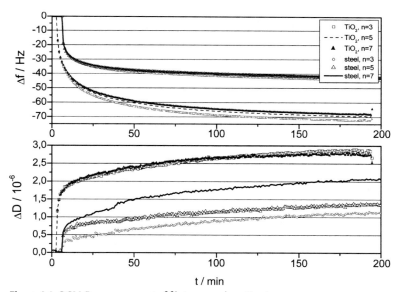

Figure 4.4 QCM-D measurement of fibrinogen adsorption to an as-delivered steel-coated quartz crystal and a sputter coated titanium oxide surface. Adsorption curves are shown for overtones 3, 5 and 7 of the basic resonance frequency.

titanium and steel surfaces at three different overtones ($n = 3, 5, 7$) of the driving frequency. The total mass of adsorbed protein was much higher on titanium than on steel, indicated by higher final Δf values (about $-70\,\text{Hz}$ compared to $-43\,\text{Hz}$). For both materials the three adsorption curves lay close to each other, thus showing no significant influence of the driving frequency. However, regarding the ΔD values, proteins showed a totally different behavior on TiO_2 compared with steel (see Figure 4.4). While the dissipation curves obtained with different frequency overtones lay close together for TiO_2 with final values of about 2.8×10^{-6}, the curves proceeded separately for the fibrinogen adsorption on steel. The dissipation increased significantly with the number of the overtone. Furthermore, the Δf and ΔD values obtained from these experiments were used for the modeling of the viscoelastic properties of the adsorbed fibrinogen film, using the mathematical fit routines included in the QSoft program package. Figure 4.5 presents the results for the modeled viscosity of the protein film vs. time. The curve obtained from the fibrinogen adsorption on TiO_2 showed a steep slope in the first few minutes and increased only insignificantly for the rest of the experiment, the maximum value of η being about $0.00285\,\text{kg}\,\text{m}^{-1}\,\text{s}^{-1}$. On the steel-coated surface the viscosity of the protein film increased with a less steep slope, reached a maximum value of about $0.01\,\text{kg}\,\text{m}^{-1}\,\text{s}^{-1}$ after approximately 75 min, and decreased slowly for the following 125 min. The significant differences in the viscosity of the fibrinogen film on the two materials support the assumption that the protein molecules underwent high denaturation on the steel surface, thus forming a more rigid film. High viscosity also leads to an increasing penetration

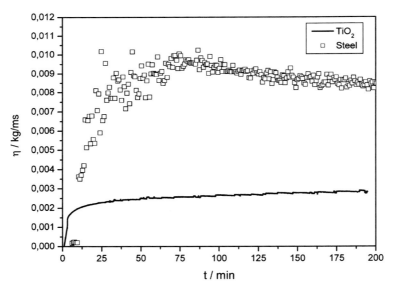

Figure 4.5 Modeled viscosity η vs. time for fibrinogen adsorption to as-delivered steel-coated and to magnetron sputtered titanium oxide layers. The fit routine used the Δf and ΔD data of all frequency overtones.

depth of the acoustic shear wave of the quartz resonator into the liquid [6], which can serve as an explanation of the strong frequency dependency of the ΔD values of the films adsorbed to the steel surface. The decrease of η after reaching a maximum value observed on the steel surface may have occurred due to a maximum coverage of the surface with denatured protein molecules and a subsequent adsorption of non-denatured molecules forming a second over-layer. Apparently, protein adsorption and film formation occurred in a much more uniform way on the TiO$_2$ surface.

Fibronectin (FN) showed similar behavior as fibrinogen, although the differences in the Δf and ΔD values were less pronounced, as can be seen in the first parts of the adsorption curves shown in Figs. 4.6 and 4.7. The curve slopes after the removal of the FN solution from the sensor chamber and the addition of BSA revealed only poor adsorption of albumin on the FN-covered steel surface and practically no adsorption on the TiO$_2$ surface, indicating the FN molecules covered the main part of the adsorption sites. The addition of the antibody P1H11 resulted in increasing Δf and ΔD values on both the TiO$_2$ and the steel surface (Figure 4.6), indicating that the cell binding domain was recognized and occupied by the antibody. The H300 antibody, specific for the C-terminal group of the FN molecule, showed a similar effect (results not shown here). However, the adsorption curve shown in Figure 4.7 reveals no adsorption of the MAB antibody, indicated by the flat line after addition of the solution. Obviously, the N-terminus of the FN molecule was not accessible for the antibody. These results were confirmed by experiments carried out in parallel using enzyme-linked immunosorbent assay (ELISA) (see Section 1.1 of Part III) and suggest the assumption that the FN molecule coupled to both the TiO$_2$ and the

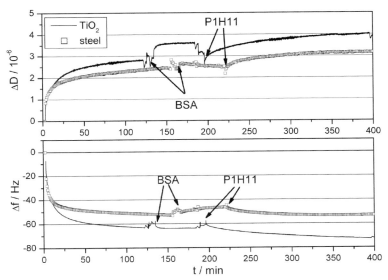

Figure 4.6 QCM-D measurement of fibronectin adsorption on TiO$_2$ and steel surfaces. Additions of BSA and P1H11 solution are marked by arrows.

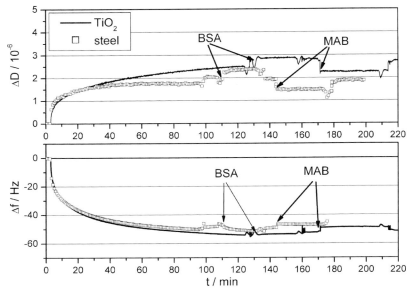

Figure 4.7 QCM-D measurement of fibronectin adsorption on TiO$_2$ and steel surfaces. Additions of BSA and MAB solution are marked by arrows.

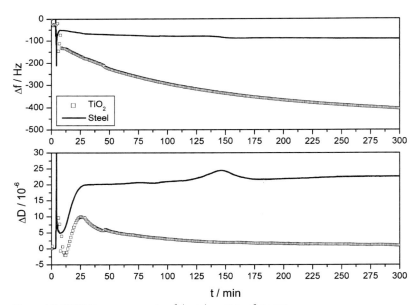

Figure 4.8 QCM-D measurements of the adsorption of mouse fibroblasts on TiO$_2$ and steel surfaces.

steel surface via its NH_4^+ group, which is consistent with the fact that both surfaces are negatively charged in solutions at physiological pH.

Finally, the adsorption of mouse fibroblasts to TiO_2 and steel surfaces was tested, using an initial concentration of 50 000 cells mL^{-1} on both samples. The experiments were carried out in culture medium under physiological conditions to enable cell growth and proliferation. The big differences in the Δf and ΔD vs. time curves, presented in Fig. 4.8, showed that the total amount of adsorbed cells was significantly higher on TiO_2 than on steel. The high dissipation of the cell film adsorbed to the steel surface was presumably due to poor cell adhesion compared to TiO_2. In addition, the ΔD curve of the latter sample showed a pronounced maximum after 25 min of adsorption, followed by a continuous decrease of ΔD. This, in combination with the simultaneous increase of the frequency shift, gave evidence that the cells started to secrete extracellular matrix, hence forming a more rigid film with a decreasing fraction of incorporated water.

The QCM-D technique has proved to be a suitable method for the time-resolved monitoring of the adsorption of proteins and subsequent desorption and adsorption processes. The semi-quantitative determination of antibody adsorption to an already adsorbed protein film provides valuable information about the orientation and conformation respectively of the protein molecules on the substrate surface. The comparative cell adsorption experiments with two different substrates have shown that QCM-D can also serve as a method for the biocompatibility evaluation of implant materials.

References

1 Höök, F., Rodahl, M., Brzezinski, P. and Kasemo, B. (1998) Energy dissipation kinetics for protein and antibody-antigen adsorption under shear oscillation on a quartz crystal microbalance. *Langmuir*, **14**, 729–734.

2 Sauerbrey, G. (1959) Verwendung von Schwingquarzen zur Wägung dünner Schichten und zur Mikrowägung. *Zeitschrift für Physik*, **155**, 206–222.

3 Voinova, M.V., Jonson, M. and Kasemo, B. (2002) 'Missing mass' effect in biosensor's QCM applications. *Biosensors Bioelectron.*, **17**, 835–841.

4 Voinova, M.V., Rodahl, M., Jonson, M. and Kasemo, B. (1999) Viscoelastic acoustic response of layered polymer films at fluid-solid interfaces: continuum mechanics approach. *Phys. Scr.*, **59**, 391–396.

5 User Manual for Q-Sense D300, QAFC 302, Q-Sense (2000).

6 McHale, G., Banerjee, M.K., Newton, M.I. and Krylov, V.V. (1999) Surface acoustic wave resonances in the spreading of viscous fluids. *Phys. Rev. B*, **59** (12), 8262–8270.

5
Computer Simulations: Modeling of Protein Adsorption on Nanostructured Metallic Surfaces

Patrick Elter and Roger Thull

Materials and Methods

In the last few years several studies have been performed in which protein adsorption on nano-patterned adsorbents was investigated and a dependence on the topography was determined [1]. A detailed understanding of the adsorption process is of great technical relevance since it opens the possibility to adjust biocompatibility by nanostructured surface modifications. In this chapter the influence of a topographical nanostructure with sharp edges and spikes is explained by a theoretical model combining Brownian Dynamics (BD) [2] and the Finite Differences (FD) [3] method. A nano-cube with an edge length of 16 nm has been selected as model system and the adsorption kinetic of hen egg white lysozyme (HEWL) is studied. HEWL, a small globular protein, is particularly suitable for the computations, since there are available both numerous experimental investigations [4,7] and previous theoretical descriptions for planar surfaces [7,9]. The theoretical description of an adsorption process on three-dimensional rough or porous nanostructures is more complex than for a planar surface: the influence of edges and spikes on the local electric field has to be considered and it may be assumed that preferred sites of adsorption are formed. An increased protein density at certain places will also affect the local neighborhood in the adsorption steps following at later times, due to protein–protein interactions. Moreover, the occupation of key positions like the top of a narrow channel by only a few proteins may exclude large areas from adsorption. Thus, it stands to reason that a theoretical prediction is required to include protein–protein and protein–nanostructure interactions as well as a calculation of the individual diffusion trajectory of each single protein.

BD simulations are an efficient way for such calculations, since the displacement and the occurring forces of each protein are calculated in small time steps and the adsorbent is gradually filled up with proteins. In the presented model, proteins are described for simplicity as uniformly charged spheres and are adsorbed from an electrolyte solution on a nanostructure. The nanostructure can be arbitrarily defined. Potential distributions of the nanostructure are calculated by the FD method and the adsorption process of the proteins is simulated by BD calculations. The electrolyte molecules (e.g. Na^+, Cl^-) are not explicitly represented, but their influences on the

Metallic Biomaterial Interfaces. Edited by J. Breme, C. J. Kirkpatrick, and R. Thull
Copyright © 2008 WILEY-VCH Verlag GmbH & Co. KGaA, Weinheim
ISBN: 978-3-527-31860-5

prevailing potential distributions are considered. In a BD simulation, the displacement of each particle is derived from the forces acting on it. The new particle position after a small time step will be considered as initial position for the next step. Thus, the gradual computation of many time steps results in the motion trajectory of a protein. The individual displacement of each protein is calculated from a systematic net force, which results from the mutual forces of the proteins along with the particle–nanostructure interactions, and a stochastic displacement. The latter has a Gaussian average distribution, a variance–covariance according to the lateral diffusion coefficient and no correlations to the systematic force. A detailed description of the algorithm for update particle positions is given in [8,10,11].

The simulation box consists of an interior region where the adsorbent is located and adsorption takes place. This region is surrounded by a protein reservoir with a constant bulk concentration, which is assumed to be far away from the nanostructure and of large dimension. Inside the interior region, space is discretized into small cubic elements with an edge length of h and a volume h^3. A partitioning of space in cubic finite elements is suitable for this calculation, because it opens the possibility for a fast calculation of the electrostatic differential equations by the FD method [3]. The total volume of the region is $N^3 h^3$, where N is a power of two and denotes the number of elements along a Cartesian axis. The nanostructure is located near the center of this region. A Boolean function is assigned to each cubic element, which is 1 if the particular element is accessible by the solute (proteins, electrolyte) or 0 if it is occupied by the nanostructure. Thus, an arbitrarily shaped nanostructure can be approximated by the combination of sufficiently small elements. Required parameters (e.g. the local charge density or the dielectric constant as shown below) can be individually assigned to each finite element. The systematic net forces are derived from the potentials and are based on the DLVO theory. They consist of an electrostatic and a dispersion part. Pairwise additivity for all interaction potentials and forces is assumed [8]. Inside the interior region, protein–protein and protein–nanostructure interactions are calculated while the protein basin is assumed to be far away and only protein–protein interactions are considered.

The electrostatic part of the protein–protein pair potential is given by the classic DLVO theory [12]:

$$U_{i,j}^{ER}(r) = \frac{q_{prot}^2 e_0^2}{4\pi\varepsilon_r\varepsilon_0} \left[\frac{\exp(\kappa a)}{1 + \kappa a}\right]^2 \frac{\exp(-\kappa r)}{r} \qquad (5.1)$$

where $\kappa^2 = (e_0^2 N_A \sum_v c_v q_v^2)/(\varepsilon_r\varepsilon_0 k_B T)$ is the inverse Debye length, a the radius of the protein sphere, $r > 2a$ the distance between two proteins, ε_0 the permittivity of free space, ε_r the relative dielectric constant, e_0 the electron charge and q_{prot} the net charge of a protein. Moreover, each ion sort v of the electrolyte is parameterized by its concentration c_v and its charge q_v. The dispersion part of two spheres of radius a can also be expressed by an analytic equation [13]:

$$U_{i,j}^{DISP} = -\frac{A}{6}\left[\frac{2a^2}{r_{i,j}^2 - (2a)^2} + \frac{2a^2}{r_{i,j}^2} + ln\frac{r_{i,j}^2 - (2a)^2}{r_{i,j}^2}\right] \qquad (5.2)$$

The protein–nanostructure interactions cannot be described by analytic equations, due to the postulation of an arbitrarily definable shape. The electrostatic part has to be calculated numerically from the nonlinear Poisson–Boltzmann equation

$$\nabla(\varepsilon_0\varepsilon_r(x)\nabla\phi) = -\varphi - \sum_\nu q_\nu e_0 c_\nu N_A \exp\left(-\frac{q_\nu e_0 \phi}{k_B T}\right) \tag{5.3}$$

by the FD method with a multigrid algorithm [3]. Here ϕ is the electrostatic potential (in this formulation in units of volts), φ the partial charge distribution belonging to the nanostructure, N_A Avogadro's number, k_B the Boltzmann constant and T the temperature of the system. The dispersion part can be obtained from a numerical integration of the Hamaker equation [13].

$$U = -\frac{A}{\pi^2} \int\limits_{V_1} dv_1 \int\limits_{V_2} dv_2 \frac{1}{r^6} \tag{5.4}$$

on a fine sub-grid where dv_1 and dv_2 designate the volume elements of the integrals over the total particle volume V_1 and V_2, respectively. Parameter r denotes the distance between dv_1 and dv_2 and A is the Hamaker constant. Short-range repulsion as known from the Lennard-Jones potential is implicitly defined by the algorithm: Each time a particular displacement leads to an overlap with another protein or the nanostructure, it is reduced until the overlap disappears. The sequence, in which this is examined, is changed every time step, in order to reduce the effect of unwanted correlations. Additionally a protein is flagged as immobilized, as soon as it grazes the surface of the nanostructure (no surface diffusion). A detailed description of the algorithm will be published elsewhere, due to its complexity and length.

HEWL can be described in good approximation by a sphere with a radius of 1.5 nm and a net charge of $+8$ at pH $=7$ [5]. The diffusion constant was estimated to 1.1×10^{-6} cm^2 s^{-1} and Hamaker constants were taken from the adsorption model of Oberholzer et al. [8] to make the results comparable to earlier calculations at planar adsorption surfaces. They were set to 2.0×10^{-20} J for the lysozyme–lysozyme interaction and 1.0×10^{-20} J for the lysozyme–nanostructure interaction. The lysozyme bulk concentration at large distances has been set to 200 µg mL^{-1} as a compromise between the usually low concentrations of atomic force microscopy measurements and the higher concentrations of bulk experiments. The nanostructure has been modeled in constant surface potential mode with a surface potential of $\phi_0 = -0.1$ V at the solid side of the interface, which roughly corresponds to the potential of mica [14]. The liquid side of the interface was configured with a sodium chloride electrolyte with concentrations between 0.0015 and 0.1 mol L^{-1} and a relative dielectric constant of $\varepsilon_r = 81$. The interior region of the model setup was discretized with a grid of 128^3 points with $h = 0.5$ nm. The time step Δt for the BD simulations has been set to 1.136×10^{-11} s, so that the average displacements are small compared to the protein radius ($\sqrt{2D\Delta t} = 0.033a$). Smaller time steps did not lead to appreciable improvements. All simulations were performed on a PC with an Intel Pentium 4 processor with 3.0 GHz and Linux operating system. The program code was developed in C++ using GNU C++ version 3.5.5.

Results and Discussion

An oppositely charged nano-cube with an edge-length of 16 nm was chosen as model system, because it was expected that the adsorption at edges plays a major role for such small structures. In this setup all protein–nanostructure interactions are attractive while the protein–protein interaction consists of an attractive dispersion and a repulsive electrostatic part. The resulting surface coverage as a function of time is displayed in Figure 5.1. Note that it has been obtained by counting the adsorbed proteins on the known surface area of the cube. This method does not lead to a maximum surface coverage of $\theta = 54.7\%$ in three dimensions as known from the planar RSA model [15], due to "surface-less" adsorption at the corners and proteins may extend into space at the edges of a finite surface. At the beginning of the adsorption process, protein–protein interactions do not play a large role and proteins are adsorbed independently with a large average distance. Hence, the surface coverage increases linearly at early time steps. At later times additional proteins can only be adsorbed with smaller distances and the repulsive Coulomb force between two protein spheres becomes apparent. As a result, the displayed coverage function deviates to smaller slopes and reaches an equilibrium value at large time steps. Equilibrium is essentially determined by the electrolyte concentration, due to an increased screening of the repulsive protein–protein Coulomb interaction at higher electrolyte concentrations. At an electrolyte concentration of $0.0015 \, mol \, L^{-1}$, repulsion outweighs any attractive contribution of the dispersion interactions and the further adsorption is prevented by an electrostatic barrier after only a few proteins were adsorbed. As the electrolyte concentration increases the electrostatic repulsion is screened and additional proteins can be inserted. At a concentration of $0.1 \, mol \, L^{-1}$ the Coulomb barrier becomes so small that the proteins are able to break through it by thermal movement. Thus, they may access distances, in which the attractive dispersion forces outweigh and the surface coverage corresponds almost to a closest

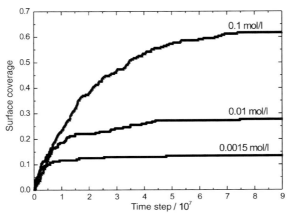

Figure 5.1 Surface coverage for the lysozyme adsorption on a $(16 \, nm)^3$ nano-cube for different NaCl concentrations.

packing. The surface coverage exhibits a greater equilibrium value than in the planar case [9] and can be explained in a more detailed investigation of the adsorption process.

Figure 5.2 displays the adsorbed proteins at the nano-cube for three different electrolyte concentrations. Figure 5.2A–C displays the adsorption locations after 500 000 (6.8 μs), 5 000 000 (68.1 μs) and 89 000 000 (1.0 ms) time steps for a salt concentration of 0.0015 mol L^{-1} NaCl. The proteins are preferentially adsorbed at the edges of the cube, which is mainly due to two different effects. First, the local electric field is higher near the edges and corners than in the center of a surface. Moreover, a point that is located at a corner or an edge can be reached by more proteins of the surrounding neighborhood than a point on a surface and already possesses a higher probability of adsorption due to geometry. Since the charged protein spheres repel each other, a higher adsorption rate at the edges leads to a lower surface coverage of the nearby planar surfaces at low salt concentrations. Only a few proteins manage to cross the repulsive barrier of the previously adsorbed proteins and reach regions where the attractive interaction of the nanostructure outweighs. At higher salt concentrations (Figure 5.2D–F for 0.01 mol L^{-1} NaCl and Figure 5.2G–I for 0.1 mol L^{-1} NaCl) the adsorption process begins in exactly the same way. However, the adsorption process can be continued at later time steps, due to the higher screening effect of the electrolyte and also the surfaces are covered. At an NaCl concentration of

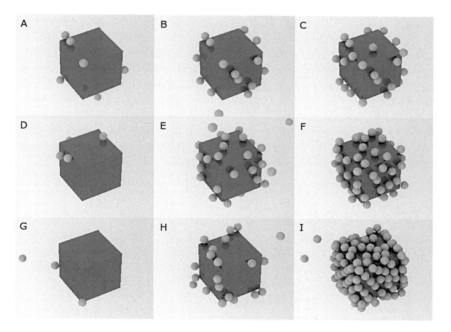

Figure 5.2 Lysozyme adsorption on a (16 nm)^3 nano-cube at time steps 500 000, 5 000 000 and 89 000 000 for NaCl concentrations of 0.0015 mol L^{-1} (A–C), 0.01 mol L^{-1} (D–F) and 0.1 mol L^{-1} (G–I).

$0.1 \, \text{mol} \, L^{-1}$ almost a closest packing is reached as already assumed on the basis of the surface coverage graph. A slight increase of the protein density at the corners of the cube is still determined. With all accomplished calculations the surface coverage never exceeds a monolayer, but proteins may extend into space at the edges and corners of the cube. Hence, the mechanism of protein adsorption is governed by two different stages. In the first part preferential places are occupied, which are determined by the electrical field near the surface and the size of the surrounding liquid volume. The dispersion forces have only a small influence on the selection of the adsorption sites in this model, due to their short range and fast raising character. If a protein reaches the regions near the surface of the nano-cube, where dispersion forces play a role, it is usually adsorbed in the immediate vicinity. In the second part of the mechanism, also more unfavorable sites are occupied. The significance of this part is determined by the electrolyte concentration and the prevailing charge ratios.

Conclusion

The model demonstrates that the adsorption behavior of proteins on a nanostructure can be basically explained in the framework of the DLVO theory and altered adsorption characteristics along the nanostructure as measured in [1] can be understood. The adsorption behavior was investigated on the basis of an oppositely charged metal cube with sharp edges and corners and an edge length of 16 nm. If the cube comes into contact with a protein/electrolyte solution, obviously first the favorable positions at the edges are occupied. At later times also the surfaces of the cube are covered, if the salt concentration is sufficient to screen the repelling protein–protein electrostatic interaction.

Since arbitrarily shaped surfaces can be calculated, this algorithm opens the possibility to make predictions about the efficiency of biomaterial surfaces concerning protein adsorption. However, the model has also some limitations: To obtain a useful accuracy the elements for the electrostatic potential calculation should not be larger than 0.5 nm and the size of the sub-grid for the dispersion energy should be smaller than 0.06 nm. For very large structures (e.g. imported AFM measurements at rough metal surfaces) the computing time is still too high. In practice the volume of the simulation box is limited therefore to $(128 \, \text{nm})^3$.

References

1 Galli, C., Coen, M.C., Hauert, R., Katanaev, V.L., Wymann, M.P., Gröning, P. and Schlapbach, L. (2001) Protein adsorption on topographically nano-structured titanium. *Surf. Sci.*, **474**, L180–L184.

2 Chen, J.C. and Kim, A.S. (2004) Brownian dynamics, molecular dynamics, and Monte Carlo modeling of collodial systems *Adv. Colloid Interf. Sci.*, **112**, 159–173.

3 Press, W.H., Flannery, B.P., Teukolsky, S. A. and Vetterling, W.T. (1993) Numerical Recipes in C, Cambridge University Press.

4 Kim, D.T., Blanch, H.W. and Radke, C.J. (2002) Direct imaging of lysozyme adsorption on mica by atomic force microscopy. *Langmuir*, **18**, 5841–5850.

5 Tanford, C. and Roxby, R. (1972) Interpretation of protein titration curves

application to lysozyme. *Biochemistry,* **11**, 2192–2198.

6 Muschol, M. and Rosenberger, F. (1995) Interactions in undersaturated and supersaturated lysozyme solutions: static and dynamic light scattering results. *J. Chem. Phys.,* **103**, 10424–10432.

7 Wahlgren, M., Arnebrant, T. and Lundstrom, I. (1995) The adsorption of lysozyme to hydrophilic silicon oxide surfaces: comparison between experimental data and models for adsorption kinetics. *J. Colloid Interf. Sci.,* **175**, 506–514.

8 Oberholzer, M.R., Wagner, N.J. and Lenhoff, A.M. (1997) Grand canonical Brownian dynamics simulation of colloidal adsorption. *J. Chem. Phys.,* **107**, 9157–9167.

9 Ravichandran, S. and Talbot, J. (2000) Mobility of adsorbed proteins: a Brownian dynamics study. *Biophys. J.,* **78**, 110–120.

10 Ermak, D.L. (1975) A computer simulation of charged particles in solution: I. Technique and equilibrium properties. *J. Chem. Phys.,* **62**, 4189–4196.

11 Van Gunsteren, W.F. and Berendsen, H.J.C. (1982) Algorithms for brownian dynamics. *Molec. Phys.,* **45**, 637–647.

12 Verwey, E.J.W. and Overbeek, J.T.G. (1948) Theory of the Stability of Lyophobic Colloids, Elsevier, Amsterdam.

13 Hamaker, H.C. (1937) The London–Van der Waals attraction between spherical particles. *Physica.,* **4**, 1058–1072.

14 Rojas, O.J., Ernstsson, M., Neumann, R.D. and Claesson, P.M. (2002) Effect of polyelectrolyte charge density on the adsorption and desorption behavior of mica. *Langmuir,* **18**, 1604–1612.

15 Schaaf, P. and Talbot, J. (1989) Surface exclusion effects in adsorption processes. *J. Chem. Phys.,* **91**, 4401–4409.

III
Biological Characterization of the Interface and Materials-Related Biosystem Reactions

1
Protein and Cell Adhesion Mechanisms

1.1
Characterization of Surface-Dependent Protein Adsorption

Andrea Ewald, Patrick Elter, and Roger Thull

Cell adhesion to implant surfaces is determined by protein adsorption to the surface, building a first material-dependent layer of extracellular matrix proteins [1]. Surface characteristics like isoelectric point (IEP) or roughness affect protein structure leaving the macromolecules native or changing their structure thus influencing the state of this layer [2]. Especially rough surfaces may lead to electric field strengths being able to change protein conformation. The matrix proteins are either provided by cell culture media and blood plasma, or synthesized by the cells.

Materials and Methods

Stainless steel according to standard 316 L and a commercially pure titanium cylinder (1 mm height, 15.5 mm diameter) were used machined and polished. In addition titanium dioxide was deposited on these surfaces using a custom-made physical vapor deposition system. Coatings were about 200 nm thick. After coating, the samples were cleaned by rinsing in 2% SDS (Sigma, Taufkirchen, Germany), pure water, 5% Extran® (Merck, Germany), pure water, isopropanol and pure water, each step being for 10 min in an ultrasonic bath at 40 °C. Cleaned samples were used immediately.

Surface Polarization and Protein Adsorption Surfaces were polarized with -600 mV and $+600$ mV respectively in a polarization chamber. Proteins were adsorbed for 30 min at 37 °C from fetal calf serum (FCS; Invitrogen, Karlsruhe, Germany) and 0.1% bovine serum albumin (BSA; Sigma, Taufkirchen, Germany) in phosphate buffered saline (PBS; 137 mM NaCl, 2.7 mM KCl, 7.0 mM Na_2HPO_4, 1.5 mM KH_2PO_4). The amount of adsorbed proteins was determined after several rinsings in PBS using the DC-protein assay (Bio Rad, Munich, modified according to [3]) according to the manufacturer's instructions.

Metallic Biomaterial Interfaces. Edited by J. Breme, C. J. Kirkpatrick, and R. Thull
Copyright © 2008 WILEY-VCH Verlag GmbH & Co. KGaA, Weinheim
ISBN: 978-3-527-31860-5

Identification of Protein Orientation   To identify protein orientation after adsorption of fibronectin, we used three primary antibodies recognizing different epitopes of the protein. These were the N-terminus (Chemicon, Hofheim/Taunus, Germany, $1\,\mu g\,mL^{-1}$), the cell binding domain (P1H11, Santa Cruz, Heidelberg, Germany, $1\,\mu g\,mL^{-1}$) and the C-terminus (H300, Santa Cruz, $1\,\mu g\,mL^{-1}$).

The enzyme-linked immunosorbent assay (ELISA) was done as described earlier [4].

Fluorescent Labeling of Adsorbed Proteins   Fibronectin ($50\,\mu g\,mL^{-1}$), collagen I (Sigma, Taufkirchen, Germany, $1\,mg\,mL^{-1}$), and BSA (Sigma, $1\,mg\,mL^{-1}$) were adsorbed to polished and machined stainless steel discs for 1 h at room temperature. After three rinsings in PBS (137 mM NaCl, 2.7 mM KCl, 7.0 mM Na_2HPO_4, 1.5 mM KH_2PO_4) the samples were incubated with primary antibodies (anti fibronectin H300, $1\,\mu g\,mL^{-1}$; anti BSA, $1\,\mu g\,mL^{-1}$; anti collagen I, Sigma, $1\,\mu g\,mL^{-1}$) followed by additional rinsing in PBS. Incubation with appropriate secondary antibodies coupled to Cy2 (Dianova, Hamburg, Germany) was done for 20 min at room temperature. The samples were rinsed again three times in PBS and air dried from ethanol. Fluorescence was quantified using a spectrofluorometer at a wavelength of 450 nm.

Results

Polarization-Dependent Protein Adsorption   To mimic electric field forces, stainless steel and titanium surfaces have been polarized with $-600\,mV$ and $+600\,mV$ respectively [5]. After 30 min protein adsorption to polarized stainless steel surfaces from FCS (Figure 1.1A) nearly the same amount of protein on negatively polarized and nonpolarized samples could be detected. The ratio of protein on polarized to nonpolarized surfaces was about 1 (Figure 1.1A). In contrast, on the positive charged surfaces the ratio was >1 indicating more protein on the polarized surfaces than on the nonpolarized ones. Protein adsorption from 0.1% BSA to stainless steel showed no polarization dependency. The ratio of protein on negatively charged surfaces to protein on uncharged control surfaces, as well as the ratio of protein on positive charged surfaces to control was about 1 (Figure 1.1C).

After protein adsorption to titanium surfaces from FCS significantly more protein was adsorbed to negatively polarized surfaces compared to control (ratio >1.25, Figure 1.1B). On positive charged titanium surfaces less protein than on control surfaces could be detected (ratio between 0.8 and 0.9, Figure 1.1B). BSA adsorption to negatively polarized titanium surfaces was below that on control surfaces (Figure 1.1D). The ratio of adsorbed BSA on positive charged titanium to adsorbed BSA on control surfaces was about 1, indicating that there was no difference of BSA adsorption to these surfaces (Figure 1.1D).

Protein Orientation Dependent on Surface Material   The orientation of fibronectin adsorbed to different surfaces was determined by means of ELISA using antibodies recognizing three different epitopes, namely the N-terminus, the C-terminus and the cell binding domain, of the protein. We used commercially pure titanium,

Protein adsorption from FCS

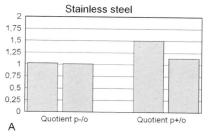

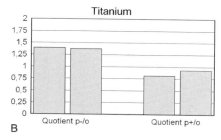

A

B

Protein adsorption from 0.1% BSA

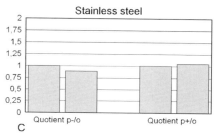

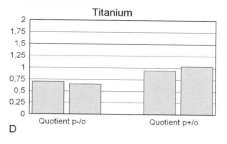

C

D

Figure 1.1 Surface polarization shows effects on protein adsorption dependent on substrate material. Surfaces have been negatively or positive polarized. The ratio of the amount of adsorbed protein on polarized surfaces to the amount of protein on nonpolarized control surfaces (o) has been calculated (p−, negatively polarized; p+, positively polarized). (A, B) Protein adsorption from FCS to stainless steel and titanium, respectively; (C, D) protein adsorption from 0.1% BSA.

polished and machined and these coated with titanium dioxide, and stainless steel according to standard 316 L, also polished, machined and coated with titanium dioxide. Figure 1.2 shows the results obtained for protein orientation on these surfaces. In all cases the antibody recognizing the N-terminus showed the weakest signal, indicating a masking of this epitope which may be due to binding of the protein to the surface via the N-terminus. The C-terminal epitope was recognized well by the antibodies, indicating good accessibility for the antibodies. Only on TiO$_2$-coated titanium surfaces (Figure 1.2C) this epitope seems to be masked. The cell binding domain of fibronectin was detected by the antibody, but to a lesser extent than the C-terminal domain. The difference in accessibility between cell binding domain and C-terminus was more pronounced on titanium surfaces than on the stainless steel surfaces.

Adsorption of Three Different Proteins to Nanostructured Surfaces To examine protein adsorption of differently charged proteins to nanostructured ($R_a = 12$ nm) and machined ($R_a = 1.34$ μm) stainless steel surfaces we incubated the samples with fibronectin, collagen I or BSA [5]. Fibronectin adsorption to the nanostructured surfaces was twice as high as on the microstructured ones (Figure 1.3). Collagen did

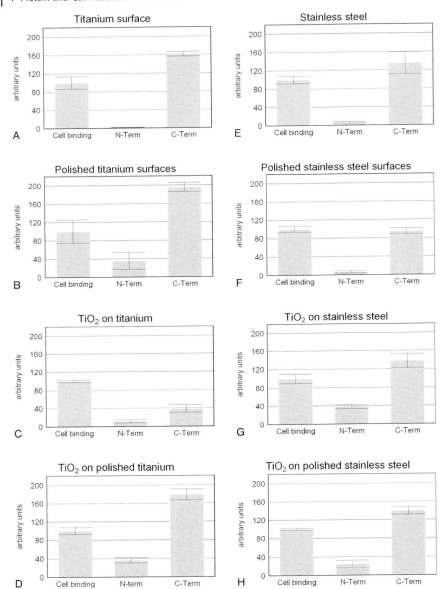

Figure 1.2 Determination of protein conformation after adsorption by means of ELISA using antibodies recognizing three different protein domains. Adsorption of fibronectin to titanium surfaces: (A) machined, (B) polished, (C) coated with TiO_2 and (D) polished and coated with TiO_2. Adsorption of fibronectin to stainless steel according to standard 316 L: (E) machined, (F) polished, (G) coated with TiO_2, and (H) polished and coated with TiO_2.

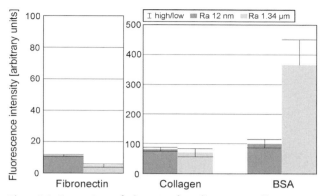

Figure 1.3 Comparison of adsorption from fibronectin, collagen and BSA to nanostructured and microstructured stainless steel surfaces after fluorescence labeling. Protein adsorption is dependent on surface roughness and protein charge.

not show any preference for nano- or microstructured surfaces, in contrast to BSA, where the amount of adsorbed protein on microstructured samples was 3.5 times the amount of protein on nanostructured samples (Figure 1.3).

Discussion

Protein Adsorption to Polarized Surfaces Biocompatibility of metallic implants is determined by the protein adsorption to these materials. After placement of the implant into a biological environment a protein film adsorbs before cell attachment occurs. Depending on protein conformation after the adsorption process, the implant will integrate properly or be rejected by giant cell formation and encapsulation [6].

Protein conformation on the surfaces is *inter alia* affected by electric field forces [2]. To mimic these forces *in vitro*, stainless steel and titanium surfaces have been polarized with −600 mV and +600 mV. For BSA no difference in the adsorbed protein amount on stainless steel depending on positive and negative polarization could be detected. In contrast protein adsorption from FCS to stainless steel was increased on positively charged surfaces. The effect leading to this increase has to be analyzed in further experiments.

On the titanium surfaces protein adsorption from FCS was significantly increased on the negatively charged surfaces. On positive charged surfaces protein adsorption tended to be decreased. FCS is a mixture of plasma proteins, and those with basic IEP may adsorb preferentially to the positive charged surfaces. In contrast, there may be less acidic proteins in FCS and therefore protein adsorption is not affected on negatively charged surfaces when compared to uncharged control surfaces. Adsorption from BSA to charged titanium surfaces yields a protein reduction on negatively charged surfaces which may be due to the electrostatic repulsion between the negatively charged protein and the identically charged surface.

Orientation of Adsorbed Protein   To determine protein orientation depending on the surface material, antibodies recognizing three different epitopes of fibronectin have been used in ELISA. On all surfaces tested the signal for the N-terminus of the protein was the weakest, indicating a binding via this part of the protein. These results were confirmed by quartz crystal microgravimetry with dissipation monitoring (QCM-D) measurements (Chapter 4 of Part II) where no antibody binding could be observed. As stainless steel surfaces are negatively charged at physiological pH [7] a binding via positively charged protein domains is comprehensible. Also on uncharged titanium and titania surfaces the binding was via the N-terminus. The C-terminal domain was well accessible for the antibodies, whereas the cell binding domain was not as intensely marked as the C-terminus. These results indicate a vertical orientation of the protein presenting the C-terminus and masking the N-terminus. Due to steric effects the cell binding domain in the middle of the protein was not completely accessible for the appropriate antibody. This orientation could be found on all surfaces tested with one exception. On titanium coated with TiO_2 the C-terminal signal was weak. Roach et al. described a vertical orientation for another filamentous protein, fibrinogen, by means of infrared spectroscopy [8].

Comparison of Adsorption of Different Proteins   The influence of nanostructured versus microstructured surfaces and the intrinsic electric field forces on the adsorption of three proteins with different IEP was analyzed by fluorescent labeling of the proteins. The adsorption behavior of fibronectin, collagen I and BSA at $pH = 7.4$ to nanostructured ($R_a = 12$ nm) and microstructured ($R_a = 1.34$ μm) stainless steel surfaces was analyzed. The IEP of fibronectin is about 6, so the protein has a slight positive charge at $pH = 7.4$. The surfaces are negatively charged (IEP of 3–5) [7] at physiological pH. We could detect two times more of the protein adsorbed on the nanostructured surfaces than on the rough ones. For collagen I adsorption the surface roughness has no influence. As the IEP of collagen is about 7–9, the protein may be scarcely charged at physiological pH. The electrostatic forces may therefore overcome the electric field forces arising due to edges on the surfaces. In contrast, BSA adsorption was affected by surface structure. On the rough surfaces 3.5 times the amount of protein than on smooth surfaces can be detected which could not be completely attributed to the surface area enlargement (which was about 3%). For BSA adsorption surface structure may play an important role, as the protein and the surface tend to show similar charges. On the nanostructured surface the repulsive electric field forces generated by edges in the material may be higher than on the microstructured surface. On the latter surface the slopes creating the roughness seem to be smooth for the small protein (about 4–5 nm in diameter). There are no additional repulsive forces generated at these slopes. For these reasons more protein could be detected on the rougher surfaces.

1.2
Integrin- and Hyaluronan-Mediated Cell Adhesion on Titanium

1.2.1
Hyaluronan-Mediated Adhesion

J.G. Barbara Nebe and Frank Lüthen

Proliferation and function of cells are controlled by integrin-mediated interactions of cells with extracellular matrix proteins like collagen or fibronectin which are synthesized and secreted by the cells [9,10]. However, little is known about the mechanism of cell adhesion during the initial phase of cell attachment to a material surface prior to the secretion of matrix proteins. Recent findings indicate a key role of the pericellular matrix substance hyaluronan (HA), which is a large linear glycosaminoglycan, in interface interactions [11–14]. The focus of our investigations was to find out the mechanism of initial cell attachment on structured pure titanium surfaces [15].

Materials and Methods
Titanium of technical purity (grade 2) was used as discs of 30 mm in diameter. The physical structure of the surface was modified by DOT GmbH Rostock using the following technologies: polishing (P), machining (NT), blasting with glass particles (GB) and blasting with corundum particles (CB) as described in detail in Section 3.1 [16]. As controls we used tissue culture plastic (TCPS) and collagen I-coated TCPS. MG-63 human osteoblastic cells were cultured in serum-free DMEM because hyaluronidase (HAdase) is inactive in serum-containing medium.

To analyze the expression of HA in osteoblasts, cells were fixed with PFA (4%), stained with biotinylated HA acid binding protein (Calbiochem) for 30 min as well as streptavidin-FITC. Fluorescence was detected by confocal microscopy (LSM 410, Carl Zeiss).

For adhesion experiments, cell spreading and actin formation cells were pre-incubated with HAdase (type IV-S, Sigma, 600 U) for 30 min (37 °C) in serum-free medium and further cultivated during the experiment under the influence of HAdase. For quantification of the initial adhesion, cells were seeded on different structured titanium surfaces and the nonadhesive cells were measured by flow cytometry. For actin formation after 3 h cells were stained with phallacidin-BodiPy after permeabilization with ice-cold acetone (Merck).

Results
The human osteoblastic cells MG-63 expressed HA, which can be cleaved by HAdase (Figure 1.4). On pure titanium surfaces with different topography, HA plays a significant role for cell attachment in the first minutes of the cell's contact to the substrate, because HAdase treatment dramatically reduced cell adhesion (Figure 1.5). On the collagen-coated surface, the effect of HAdase was not as pronounced compared to HAdase-treated cells on titanium surfaces, as well as on TCPS (*t*-test, $p < 0.01$), because osteoblasts partially compensate for the lack of HA via their integrin receptors, which bind to the collagen I. Also for cell spreading processes HA

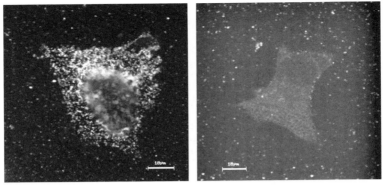

Figure 1.4 Hyaluronan (HA) expression in human MG-63 osteoblasts (left), and treatment with hyaluronidase which cleaves HA (right). Confocal microscopy LSM 410.

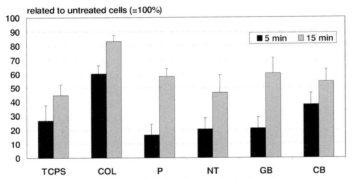

Figure 1.5 Hyaluronidase significantly inhibited the initial adhesion (5, 15 min) of MG-63 osteoblasts on structured titanium surfaces as well as on the controls, compared to untreated cells = 100%. TCPS = tissue culture plastics, COL = collagen I, P = Ti polished, NT = Ti machined, GB = Ti glass particle blasted, CB = Ti corundum particle blasted; t-test, paired: 5 min: $p < 0.05$, 15 min: $p < 0.01$.

seems to be the mediator because HAdase treatment of osteoblasts resulted in a reduced cell area as seen in scanning electron microscopy (SEM) images (Figure 1.6). Because of decelerated initial adhesion and spreading, the actin formation is also impaired in HAdase-treated cells (Figure 1.7). This influence is clearly to be seen on titanium where actin did not appear in small stress fibers as seen in cells on collagen at the same growth time of 3 h.

Discussion

The initial adhesion of osteoblasts on the structured pure titanium surface – i.e. the first contact of the cell to the substrate – is mediated by HA which is a negatively

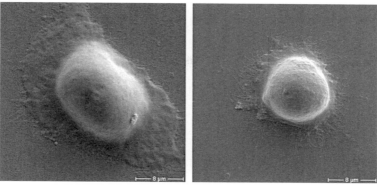

Figure 1.6 Hyaluronan mediates cell spreading – the human osteoblast (MG-63) is well spread after 30 min of attachment (left) whereas a cell treated with HAdase has a dramatically reduced cell area (right). SEM image.

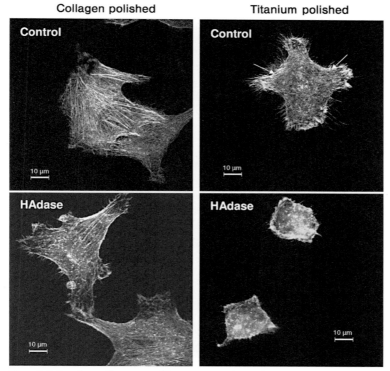

Figure 1.7 Decelerated formation of the actin cytoskeleton after 3 h in HAdase-treated MG-63 osteoblasts. Especially on the titanium surface this influence is impressive: the actin cytoskeleton of normal cells starts to develop filaments after 3 h (arrow), whereas no filaments can be observed in HAdase-treated cells. In addition, the cell area is clearly reduced. Confocal microscopy LSM 410.

charged molecule. Therefore, we conclude that a design of a biomaterial surface with positively charged chemical groups could be sufficient for mediating this first cell adhesion process. Our current investigations reveal that amino-functionalization of pure titanium [17,18] significantly improves the initial steps of the cellular contact to this metal surface, like adhesion and spreading, as well as assists gene expression of differentiation proteins on the mRNA level [19,20].

1.2.2
Integrin-Mediated Adhesion on Nanostructures

Thomas K. Monsees and Richard H.W. Funk

Osteoblast adhesion depends on specific surface properties, e.g. chemistry and structure. Cells do react to an artificial surface topography on the micrometer scale [21–23] but, as recently demonstrated, also to nanostructures [24–28]. Focal contacts are formed at the leading edge of a migrating cell and are the closest contacts between cell and implant surface. They are associated with the cytoplasmic domain of transmembrane integrin receptors on the one side and with actin stress fibers on the other. Vinculin is one of the several proteins involved in the complex of a focal contact. To analyze the effect of nanostructures on cell adhesion and focal contacts, we used patterns of parallel titanium oxide lines. These structures were produced by laser-induced oxidation of titanium thin films on glass discs and have different widths (0.2–10 μm) and distances (2–20 μm, 1000 μm), but a common height of only 12 nm (see Section 3.4 of Part I for details).

Materials and Methods
Human osteogenic sarcoma cells (SaOS-2; ATCC HTB 85) were cultured in 85% McCoy's 5A medium containing 15% fetal calf serum at 37 °C and 5% CO_2. The sterilized titanium discs were placed into culture wells and carefully covered with SaOS-2 cells at a density of 10 000 cells cm^{-2}. Actin fibers and vinculin protein were visualized after 2 days by staining with TRITC-conjugated phalloidin and a mouse monoclonal antibody against human vinculin followed by a FITC-coupled goat anti-mouse secondary antibody, respectively. Cell nuclei were stained using DAPI (for details see [29]).

Results
We found the majority of the focal contacts placed on the oxide lines (Figure 1.8A, white arrows), whereas only a few were located between them (yellow arrows). Occasionally, larger focal contacts or focal adhesions formed bridges between two oxide lines (red arrows). The portion of focal contacts bridging two oxide lines depends on the distance between the lines. Most of them were observed on the 2 μm distance structure, some on the 5 μm, but they were rare on the 10 μm structured surface. The shape of the focal contacts seems to be altered when SaOS-2 cells grew on the structured surface. On smooth titanium most focal

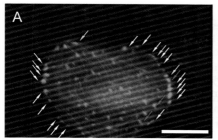 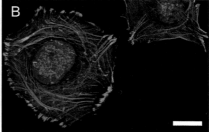

Figure 1.8 Effect of titanium surface structure on alignment and focal contact formation of SaOS-2 cells. Cells were plated for two days on either structured (parallel titanium oxide lines, height 12 nm; width 0.2 μm, distance 2 μm) or smooth (B) Ti surfaces and then fixed and triple fluorescence-stained for vinculin (green), actin (red) and nuclei (blue). Titanium oxide lines were visualized by brightfield microscopy and appeared in white. Arrows mark focal contacts located directly on titanium oxide lines (white), between oxide lines (yellow), or bridging two oxide lines (red). Bar = 40 μm (A) or 20 μm (B).

contacts were streak-like, but they appeared much rounder on the oxide lines. Thus, form and alignment of focal contacts were influenced by the topographic structures. In contrast, cells cultured on smooth titanium formed focal contacts and filopodia with no preferred orientations (Figure 1.8B). These results indicate that form and alignment of focal contacts were influenced by the topographic structures of the titanium implant.

Discussion

Our experiments using patterns of parallel titanium oxide lines with different widths (0.2–10 μm) and distances (2–20 μm, 1000 μm), but a common height of only 12 nm, indicate that form and alignment of focal contacts were influenced by the topographic structures of the titanium implant [29]. The smallest titanium oxide line width used in our study was 0.2 μm. The finding that most of the focal contacts were located on top of these lines is in accordance with Lehnert et al. [30], who demonstrated that cells adhere to and spread on extracellular matrix-coated dots as small as 0.1 μm². Teixeira et al. [24] showed corresponding data using silicone with ridges of 70 nm width and 150 nm depth. Here, corneal epithelial cells located focal contacts on the top of the ridges and only occasionally spanning the groove. Cells cultured on a smooth silicone surface, however, formed focal contacts with no preferred orientation, whereas on the nanopatterned substrates focal contacts aligned along the topographic structures. Moreover, the width of the focal contacts was dictated by the width of the underlying ridges. Dalby et al. [31] also found focal contacts on top of the nanostructures to be smaller than those on the flat control. In summary, topographic features in the nanometer range may influence orientation and shape of focal contacts by contact guidance of the osteoblast.

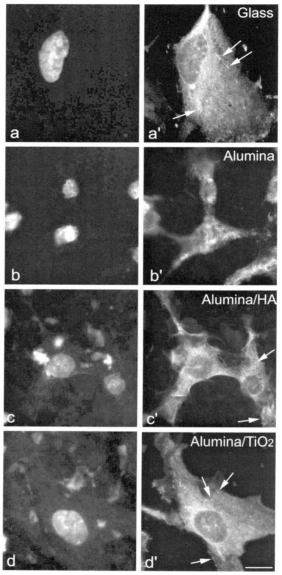

Figure 1.9 Immunofluorescence microscopy after staining with anti-vinculin (a'–d'). Cells grown on glass (a, a') as well as on titania-treated titanium (d, d') show the typical vinculin staining pattern (vinculin in focal adhesions is indicated by arrows). In contrast, cells cultivated on alumina-blasted titanium (b, b') show a different morphology with an atypical signal. On hydroxyapatite (HA)-treated titanium (c, c') cell growth is comparable to the control (focal adhesions are indicated by arrows). Corresponding DNA staining is shown (a–d). Scale bar: 15 μm.

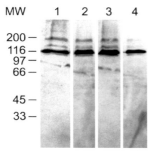

| MW | 1 | 2 | 3 | 4 |

200 —
116 —
97 —
66 —

45 —

33 —

Figure 1.10 Immunoblot analysis of cell lysate using anti-vinculin antibodies. Cells grown on alumina-blasted (lane 2) and hydroxyapatite/alumina-blasted (lane 3) titanium show a higher amount of vinculin than cells grown on control surfaces (lane 1) and titania/alumina-blasted titanium (lane 4). Molecular weight markers (MW, in kDa) are indicated.

alumina was comparable to the control. Also the immunoblot signal of vinculin from cells grown on these surfaces was not significantly increased. This indicates that the titania coating, and to a certain extent the hydroxyapatite coating, of alumina-blasting material reduces the negative effect of the alumina on the cells.

References

1 Lee, M.H., Ducheyene, P., Lynch, L., Boettiger, D. and Composto, R.J. (2006) Effect of biomaterial surface properties on fibronectin-$\alpha 5\beta 1$ integrin interaction and cellular attachment. *Biomaterials*, **27**, 1907–1916.

2 Thull, R. (2002) Physicochemical principles of tissue material interactions. *Biomol. Eng.*, **19**, 43–50.

3 Lowry, O.H., Rosebrough, N.J., Farr, A.L. and Randall, R.J. (1951) Protein measurement with the folin phenol reagent. *J. Biol. Chem.*, **193**, 265–275.

4 Hautanen, A. and Linder, E. (1981) C3c-binding ELISA for the detection of immunoconglutinins and immuno-globulin aggregates. *Method Enzymol.*, **74**, 588–607.

5 Ewald, A. and Thull, R. (2006) Surface structure/electric field dependent protein adsorption. *Biomaterialien*, **3**, 146.

6 Ratner, B.D. (2001) Titanium in Medicine (eds D.M. Brunette, P. Tengvall, M. Textor and P. Thomsen), Springer-Verlag, Berlin/Heidelberg. Chapter 1.

7 Boulangé-Petermann, L., Doren, A., Baroux, B. and Bellon-Fontaine, M.-N. (1995) Zeta potential measurements on passive metals. *J. Colloid Interf. Sci.*, **171**, 179–186.

8 Roach, P., Farrar, D. and Perry, C.C. (2006) Surface tailoring for controlled protein adsorption: effect of topography at the nanometer scale and chemistry. *J. Am. Chem. Soc.*, **128**, 3939–3945.

9 Hynes, R.O. (1992) Integrins: versatility, modulation and signaling in cell adhesion. *Cell*, **69**, 11–25.

10 Giancotti, F.G. and E Ruoslahti, E. (1999) Integrin signaling. Review signal transduction. *Science*, **285**, 1028–1033.

11 Zimmermann, E., Geiger, B. and Addadi, L. (2002) Initial stages of cell-matrix adhesion can be mediated and modulated by cell-surface hyaluronan. *Biophys. J.*, **82**, 1848–1857.

12 Cohen, M., Klein, E., Geiger, B. and Addadi, L. (2003) Organization and adhesive properties of the hyaluronan pericellular coat of chondrocytes and epithelial cells. *Biophys. J.*, **85**, 1996–2005.

13 Cohen, M., Kam, Z., Addadi, L. and Geiger, B. (2006) Dynamic study of the transition from hyaluronan- to integrin-mediated adhesion in chondrocytes. *EMBO J.*, **25**, 302–311.

14 Zaidel-Bar, R., Cohen, M., Addadi, L. and Geiger, B. (2004) Hierarchical assembly of cell–matrix adhesion complexes. *Biochem. Soc. Trans.*, **32**, 416–420.

15 Nebe, B., Lüthen, F., Diener, A., Becker, P., Lange, R., Beck, U., Neumann, H.G. and Rychly, J. (2004) Mechanisms of cell adhesion on surfaces of pure titanium. *BIOmaterialien*, **5**, 32–33.

16 Lüthen, F., Lange, R., Becker, P., Rychly, J., Beck, U. and Nebe, B. (2005) The influence of surface roughness of titanium on β1- and β3-integrin adhesion and the organization of fibronectin in human osteoblastic cells. *Biomaterials*, **26**, 2423–2440.

17 Meyer-Plath, A.A., Schröder, K., Finke, B. and Ohl, A. (2003) Current trends in biomaterial surface functionalization: nitrogen-containing plasma assisted processes with enhanced selectivity. *Vacuum*, **71**, 391–406.

18 Finke, B., Schröder, L., Lüthen, F., Nebe, B., Rychly, J. and Ohl, A. (2006) Chemical functionalization of titanium surfaces by plasma assisted polymerization. *BIOmaterialien*, **S1**, 69.

19 Lüthen, F., Schröder, K., Steffen, H., Rychly, J., Ohl, A. and Nebe, B. (2005) High initial adhesion and spreading of osteoblasts on amino-functionalized polished titanium. *BIOmaterialien*, **3**, 193.

20 Nebe, B., Lüthen, F., Finke, B., Bergemann, C., Schröder, K., Rychly, J., Liefeith, K. and Ohl, A. (2006) Improved initial osteoblast's functions on amino-functionalized titanium surfaces, *Biomol. Eng.* (submitted).

21 Brunette, D.M. and Cheroudi, B. (1999) The effects of the surface topography of micromachined titanium substrata on cell behavior *in vitro* and *in vivo. J. Biomech. Eng.*, **121**, 49–57.

22 Soboyejo, W.O., Nemetski, B., Allameh, S., Marcantonio, N., Mercer, C. and Ricci, J. (2002) Interactions between MC3T3-E1 cells and textured Ti6Al4V surfaces. *J. Biomed. Mater. Res. A*, **62**, 56–72.

23 Lu, X. and Leng, Y. (2003) Quantitative analysis of osteoblast behavior on microgrooved hydroxyapatite and titanium substrata. *J. Biomed. Mater. Res. A*, **66**, 677–687.

24 Teixeira, I., Abrahams, G.A., Bertics, P.J., Murphy, C.J. and Nealey, P.F. (2003) Epithelial contact guidance on well-defined micro- and nano-structured substrates. *J. Cell Sci.*, **116**, 1881–1892.

25 Curtis, A.S., Gadegaard, N., Dalby, M.J., Riehle, M.O., Wilkinson, C.D. and Aitchison, G. (2004) Cells react to nanoscale order and symmetry in their surroundings. *IEEE Trans. Nanobiosci.*, **3**, 61–65.

26 Martines, E., McGhee, K., Wilkinson, C. and Curtis, A. (2004) A parallel-plate flow chamber to study initial cell adhesion on a nanofeatured surface. *IEEE Trans. Nanobiosci.*, **3**, 90–95.

27 Dalby, M.J., Yarwood, S.J., Riehle, M.O., Johnstone, H.J.H., Affrossman, S. and Curtis, A.S.G. (2002) Increasing fibroblast response to materials using

nanotopography: morphological and genetic measurements of cell response to 13 nm high polymer demixed islands. *Exp. Cell Res.*, **276**, 1–9.

28 Dalby, M.J., Riehle, M.O., Sutherland, D.S., Agheli, H. and Curtis, A.S.G. (2005) Morphological and microarray analysis of human fibroblasts cultured on nanocolumns produced by colloidal lithography. *Eur. Cells Mater.*, **9**, 1–8.

29 Monsees, T.K., Barth, K., Tippelt, S., Heidel, K., Gorbunov, A., Pompe, W. and Funk, R.H.W. (2005) Effects of different titanium alloys and nanosize surface patterning on adhesion, differentiation, and orientation of osteoblast-like cells. *Cells Tissues Organs*, **180**, 81–95.

30 Lehnert, D., Wehrle-Haller, B., David, C., Weiland, U., Ballestrem, C., Imhof, B.A. and Bastmeyer, M. (2003) Cell behaviour on micropatterned substrata: limits of extracellular matrix geometry for spreading and adhesion. *J. Cell Sci.*, **117**, 41–52.

31 Dalby, M.J., Riehle, M.O., Johnstone, H., Affrossman, S. and Curtis, A.S.G. (2004) Investigating the limits of filopodial sensing: a brief report using SEM to image the interaction between 10 nm high nano-topography and fibroblast filopodia. *Cell Biol. Int.*, **28**, 229–236.

32 Gbureck, U., Masten, A., Probst, J. and Thull, R. (2003) Tribochemical structuring and coating of implant metal surfaces with titanium oxide and hydroxyapatite layers. *Mater. Sci. Eng.*, **C23**, 461–465.

33 Harris, S.A., Enger, R.J., Riggs, B.L. and Spelsberg, T.C. (1995) Development and characterization of a conditionally immortalized human fetal osteoblastic cell line. *J. Bone Miner. Res.*, **10**, 178–186.

34 Dabauvalle, M.-C., Müller, E., Ewald, A., Kress, W., Krohne, G. and Müller, C.R. (1999) Distribution of Emerin during the cell cycle. *Eur. J. Cell Biol.*, **78**, 749–756.

35 Zreiqat, H., Howlett, C.R., Zannettino, A., Evans, P., Schulze-Tanzil, G., Knabe, C. and Shakibaei, M. (2002) Mechanisms of magnesium-stimulated adhesion of osteoblastic cells to commonly used orthopaedic implants. *J. Biomed. Mater. Res.*, **62**, 175–184.

36 Thomas, J.O. and Kornberg, R.D. (1975) An octamer of histones in chromatin and free in solution. *Proc. Natl Acad. Sci. USA*, **72**, 2626–2630.

37 Khyse-Andreson, J. (1984) Electroblotting of multiple gels: a simple apparatus without buffer tank for rapid transfer of proteins from polyacrylamide to nitrocellulose. *J. Biochem. Biophys. Methods*, **10**, 203–209.

2
Material-Induced Cellular Interactions

2.1
Cell-Specific Compatibility with Materials

2.1.1
Influence of the Material Interface on Cell Migration

Thomas K. Monsees and Richard H.W. Funk

Experiments using patterns of parallel titanium oxide lines with different distances (2–20 μm), but a common height of only 12 nm [1] indicated that form and alignment of bone cells and focal contacts were influenced by the topographic structures of the titanium implant. Due to the different physicochemical properties of the Ti/Ti oxide surface used in this study, mechanisms other than mechanical guidance may also be responsible for the alignment of the osteoblast-like cells, e.g. differences in surface charges or electrical fields produced at the interface of the materials. A direct measurement of the potential difference at the semiconductor titanium oxide/metal titanium interface is difficult, but can be expected [1–3]. Therefore we used a model surface consisting of a regular pattern of gold stripes (width and distance 500 μm each, 40 nm thick; W. Pompe, M. Gelinsky, Department of Material Sciences, TU Dresden) on a titanium layer, both sputtered on glass. The interface between titanium and gold was very flat, thus any influence by mechanical contact guiding should be negligible.

Materials and Methods
Human osteogenic sarcoma cells (SaOS-2; ATCC HTB 85) were cultured in McCoy's 5A medium containing 15% fetal calf serum. Sterilized Ti–Au discs were placed into culture wells and carefully covered with SaOS-2 cells. The cell concentration (10 000 cells cm^{-2}) was very low to exclude possible influences on cellular orientation by contact with other cells. In some experiments, the inhibitor nocodazole (2 μM) was added 30 min after plating. Cells were fixed 4 h or 24 h after plating and fluorescent-stained for actin, vinculin and nuclei as described in Section 1.2. Photos of the whole Ti–Au surface pattern were divided into 500 μm thick sections (corridors), starting

Metallic Biomaterial Interfaces. Edited by J. Breme, C. J. Kirkpatrick, and R. Thull
Copyright © 2008 WILEY-VCH Verlag GmbH & Co. KGaA, Weinheim
ISBN: 978-3-527-31860-5

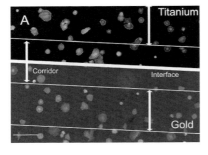

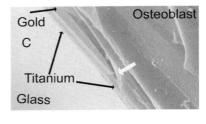

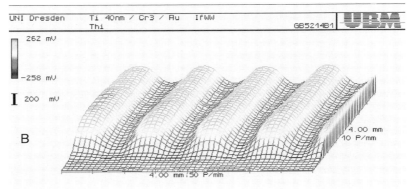

Figure 2.1 Fluorescence image of SaOS-2 cells at the Ti–Au interface (thick line). Three corridors (interface, titanium, gold) used for cell counting are indicated by white lines. Cells were fixed 4 h after plating and triple fluorescence-stained for vinculin, actin and nuclei. Scale bar = 200 μm. Raster Kelvin probe image of a structured titanium (ground)–gold (elevation) surface indicating an electrical potential of more than 150 mV at the interface. Scanning electron microscopy image of a SaOS-2 cell (C) crossing the Ti–Au interface (white arrow). (A) Glass ground; middle: titanium layer (spanning from top left to bottom right) + gold layer (B).

at the center of one of the pure metal surfaces (Figure 2.1A). Total cell numbers but also that portion of cells inside the different sections (interface, titanium or gold) with a "free view to the interface" (i.e. they can migrate without hindrance by other cells) were counted after 4 h and 24 h, respectively. Before scanning electron microscopy (SEM) imaging, cells were fixed with formalin. Thereafter, discs were critical point dried and sputtered with gold.

Results

Measurements using a raster Kelvin probe indeed demonstrated differences in the electric potential up to 150 mV at the Ti–Au interface (Figure 2.1B; M. Thieme, Department of Material Sciences, TU Dresden). Here the gold surface displays more positive surface charges whereas titanium is more negatively charged. SEM analysis shows that the topographic change between the two metals is very smooth, i.e. there is no physical hindrance for a cell to cross the interface (Figure 2.1C).

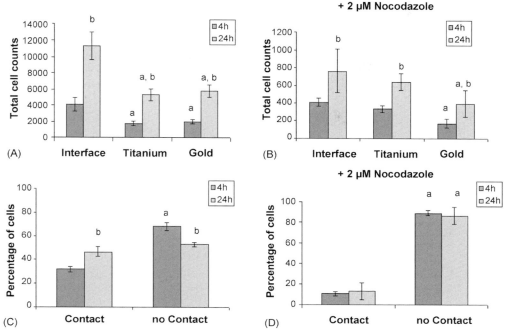

Figure 2.2 Total numbers of SaOS-2 cells in the different interface corridors, titanium or gold, 4 h and 24 h after seeding, respectively (A, B: $P < 0.05$: a vs. interface, b vs. 4 h). Percentage of cells that in principle can migrate to the interface (i.e. there is no hindrance by other cells). Shown are the percentages of cells with or without direct contact to the border of the two metals (C, D: $P < 0.05$: a vs. contact, b vs. 4 h). In some experiments (B, D) the inhibitor of microtubule dynamics, nocodazole (2 μM), was added 30 min after plating the cells.

However, the number of cells in the interface section is always (i.e. after 4 h and 24 h) significantly two times higher than in equal corridors of pure titanium or gold (Figure 2.2A). Similarly, inside the interface section, that percentage of cells with direct contact to the border of the different metals significantly increased, whereas that percentage with no contact dropped (Figure 2.2C). The ratio 24 h/4 h in cell counts is very similar within the three sections, indicating an equal proliferation rate. There is a trend to somewhat higher cell counts on the gold surface compared to titanium. These data suggest a positive cue of the interface in stimulating migration of the osteoblast-like cells from the pure metal surfaces towards the Ti–Au border.

Nocodazole disturbs microtubule dynamics, thus inhibiting cell locomotion and proliferation. This tubule de-polymerization is followed by induction of actin cytoskeleton rearrangement and formation of focal contacts [4,5]. After nocodazole treatment, that percentage of cells with direct contact to the interface dramatically dropped (Figure 2.2D). In addition, within 20 h, the increase of that cell portion with contact and the corresponding decrease in the portion without contact is now negligible. The total cell numbers in the three corridors still show significant differences between 4 h and 24 h after plating (Figure 2.2B). However, the pattern between the different corridors has changed. There is still a significant difference in total cell numbers between the corridors of interface and gold, but no more between interface and titanium. Furthermore, the titanium surface has now higher cell numbers than that of gold. The ratio 24 h/4 h in cell counts is very similar in the three sections and also very close to that without addition of nocodazole (Figure 2.2A). This may indicate a slower rate of cell division compared to the control experiment but nearly analogous proliferation rates in the different corridors.

In summary, the data show an active migration of SaOS-2 osteoblast-like cells from the pure metal surfaces towards the Ti–Au interface. Because of the smooth topographic change at the interface, contact guidance may be negligible. However, a likely mechanism for this positive migration cue may be the difference in the electrical potential measured at this interface. Such static electric fields are known to cause movement and guiding of various cell types. They also exist in mammalians causing an electrotaxis response as shown in healing wounds, developing embryos and lens epithelial cells (reviewed in [6]). Our present data show that the migration pattern is altered after blocking of microtubule dynamics using nocodazole, indicating that cells may move more from the titanium side towards the interface than from the gold surface. Strength and kinetics of adhesion are very similar on pure titanium and gold surfaces (Th. K. Monsees, unpublished). However, due to the generated electrical potential at the Ti–Au interface the gold implant has the more positively charged surface. This may attract the osteoblasts with their mainly negatively charged cell membrane. Also, the titanium implant displays more negative surface charges at the interface, which may propel the cells towards the gold implant.

Similar differences in surface potential and resulting charges may also occur at the titanium natural oxide/laser-induced oxide interface lines. These gradients seem to attract the osteoblast-like cells to migrate, to align towards and to make new focal contacts on the lines. Transmission profile analysis indicates that the maximum transparency of the laser-induced titanium oxide line is 61%, whereas the initial

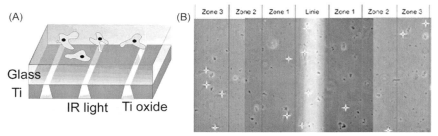

Figure 2.3 Use of the structured titanium surface as an optical mask. Cells were seeded on the fibronectin-coated back of the disc and illuminated through the structured mask with infrared (IR) light (875 nm). Phase contrast microscopy of the structured titanium disc after 3 h of irradiation. Width of the laser-induced titanium oxide lines is 80 μm, distance is about 1000 μm. Fibroblasts adhere preferentially above the light-transmitting titanium oxide lines.

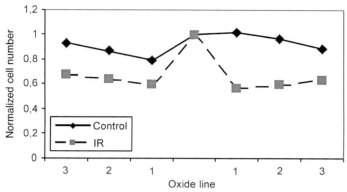

Figure 2.4 Cell numbers were counted after 3 h of illumination with infrared (IR) light in 7 corridors of equal distance. A significant amount of cells accumulate above the titanium oxide lines, whereas cell numbers in neighboring corridors dropped.
In controls, which were not illuminated, no significant variations in cell numbers between the different corridors occurred.

transparency of the titanium film was only 6.3% [7]. This optical 1 : 10 contrast clearly indicates differences in physicochemical properties between titanium surface and titanium oxide lines and may be exploited for use as an optical mask. One example is shown in Figure 2.3A: here 3T3 fibroblasts are seeded on the fibronectin-coated backside of the structured titanium disc, whereas infrared light (875 nm) illuminates through the titanium oxide pattern [8]. After 3 h, the fibroblasts accumulate mainly above the light-transmitting oxide lines (Figure 2.3B). Counting the cells in 7 corridors of equal distance also demonstrates a significant accumulation of cells above the oxide lines. In contrast, cell numbers in the darker neighboring corridors decrease (Figure 2.4). In controls, that are not illuminated, no significant variations in cell numbers between the different corridors occur. This experiment demonstrates that the fibroblasts move from the darker areas (shielded by the titanium mask) towards the light source (the very transparent titanium oxide lines). Thus, near-infrared light may induce cell migration, a phenomenon called positive phototaxis [9].

Discussion
Our data using structured implants with the metals titanium and gold demonstrate that electrical potentials do arise at the interface. This will cause differences in the surface charge, which may induce the observed active migration of SaOS-2 osteoblast-like cells from the pure metal surfaces towards the Ti–Au interface. Similar differences in surface potential and resulting charges may also occur at the titanium natural oxide/laser-induced oxide lines interface (see Section 3.2). These gradients seem to attract osteoblasts to migrate, to align towards and to make new focal contacts on the lines [1]. Such weak static electric fields (DC electric fields) are known to cause movement and guiding of various cell types, including bone cells, *in vivo* and *in vitro*.

Endogenous electric field gradients causing an electrotaxis response were shown in healing wounds, developing embryos and lens epithelial cells (reviewed in [6]). Similarly, coatings that alter the charge or the spatial arrangement of charges of an implant surface are known to improve adhesion and to induce cellular orientation and proliferation. Chondroitin sulfate proteoglycans and glycosaminoglycans, for example, are components of the embryonic extracellular matrix (ECM) and markedly enhance cell adhesion and function on titanium alloys [10]. They contain charged sulfate and carboxyl groups in a specific three-dimensional orientation and markedly enhance the orientation of nerve cells in a DC electric field by interacting with charged membrane proteins [11]. Also, hydroxyapatite ceramics with modified surface charges affect adhesion, growth and orientation of different cells types, including bone cells [12].

2.1.2
Titanium–Cell Interface: Cell Biology and Growth Behavior

Hartwig Wolburg, Friederike Pfeiffer, Andreas Heeren, Wolfgang Henschel, Jürgen Geis-Gerstorfer, Lutz Scheideler, and Dieter Kern

In an interdisciplinary approach we tried to characterize the interaction between cultured cells and an artificial substrate. To this end, we etched at a micrometer scale a geometrical texture into the surface of silicon wafers. We replicated this texture by the artificial resin Araldite and coated this resin bloc with 50 nm titanium, seeded cells on it and embedded the whole sandwich in another Araldite or Lowicryl for electron microscopy investigations. Alternatively, we grew cells directly on microstructured silicon wafers sputtered with titanium and observed the expression of molecules which would be relevant for the attachment or spreading of cells on artificial substrates. Among these molecules were the cytoskeletal protein actin, the cytoskeleton–membrane linker molecule vinculin and members of the ERM family ezrin, radixin and moesin which are thought to be responsible for the induction and regulation of structures determining the cellular surface topography such as lamellopodia, microvilli or other enlargements of the surface. We were able to show that the immunoreactivity against radixin, but not ezrin and moesin, increased during growth of gingival fibroblasts on microstructured surfaces. Furthermore, in first shear stress experiments it became clear that the material titanium itself – independent of its surface structure – seemed to mediate stronger adhesiveness than other substrate materials such as glass or plastics. Again, radixin was the molecule which gave the strongest immunoreactivity in cells growing on titanium rather than on other surfaces.

In implantology the direct effects of metallic surface structures play a decisive role for the success of integration of the implant into the tissue. These effects can be modulated by the expression of surface or adhesion molecules and can therefore be understood as a specific responsive behavior on changing properties of the implant surface [13]. In order to investigate interactive processes between cell and substrate, micro- and nanostructured artificial surfaces can be used, which in our case were resin replicas sputtered with a thin layer of 10–50 nm titanium (for detailed description of

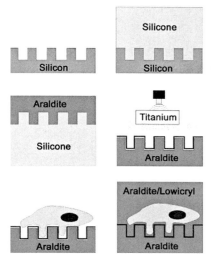

Figure 2.5 Scheme of experimental procedure. Production of a silicon wafer with microstructures, replica in silicone, re-replica in Araldite, polymerization of Araldite and sputtering with titanium, seeding of the titanium surface with cells and embedding of the whole sandwich in Araldite or Lowicryl.

the method, see [14–16]; Figures 2.5 and 2.6) or titanium-coated structured silicon wafers.

The basic processing steps for the production of the substrates with structures measuring >1 μm include the fabrication of a photomask with gratings of different linewidths and spacings by electron beam lithography, photolithography and chrome liftoff to generate an etch mask and transfer of the grating structures into a silicon substrate by a dry etching process using a CF_4/O_2 plasma [17]. By this process grooves and ridges were produced revealing a width of 1, 2, 3, 5 and 10 μm and a depth of 0.4, 0.8, 1.6 and 2 μm (Figure 2.6C).

Two possibilities arose to apply the produced structures. First, the structures could be sputter-coated with titanium in order to serve as a substrate for cell biological experiments. Second, the silicon substrate could be used as a master form for the production of a silicone replica, which in turn was the basis for the production of a secondary substrate consisting of the resin Araldite (Figure 2.6A). In order to produce nanostructured surfaces grating structures were defined by means of electron beam lithography and consecutive development using a SiO_2-like negative-type resist system (HSQ) [18] on top of a silicon wafer. The period of the grating was 200 nm, the height of the lines of the order of 100 nm was defined by the thickness of the resist. The lithography was followed by a sputter process by which the surface was coated with a 10 nm titanium layer. A typical example is shown in Figure 2.7. In this case, the substrates were directly used for biological tests.

For this aim, cells were seeded on the titanium-coated substrates (Figures 2.5 and 2.8). The investigation of these cells was done either by scanning electron microscopy

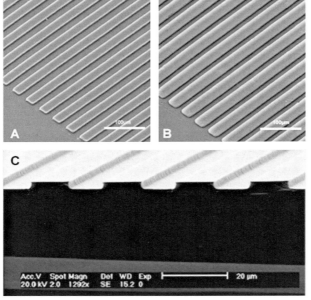

Figure 2.6 SEM micrographs of silicon master and of an appropriate Araldite replica with a microstructure (20 μm period and 2 μm high ridges), and of a microstructured silicon probe sputtered with 100 nm titanium layer (20 μm period and 2 μm depth of etching).

(Figure 2.8), fluorescence microscopy (Figures 2.9 and 2.11), confocal laser scanning microscopy including immunocytochemical staining of different antigens (Figure 2.10) or by a specially modified transmission electron microscopy. In this latter case, the microstructured silicon lattices were replicated with silicone and this again replicated with the resin Araldite. After polymerization, this Araldite lattice was sputtered with titanium, seeded with cells and after fixation the whole sandwich was embedded in another Araldite or – dependent on the question asked – Lowicryl in

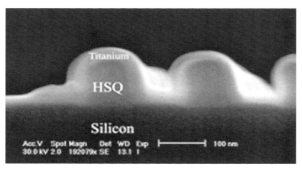

Figure 2.7 SEM micrograph of HSQ lines, which were coated with 10 nm titanium.

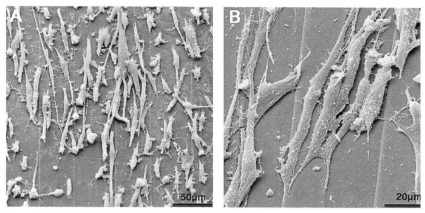

Figure 2.8 SEM micrographs of cells grown on titanium-coated microstructures.

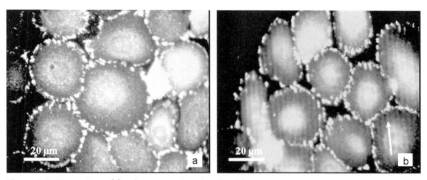

Figure 2.9 Formation of focal contacts (as visualized by an antibody against vinculin) and orientation of human keratinocytes on (a) a smooth surface and (b) a 3 μm period structure (the arrow indicates the orientation of grooves).

order to be able to label proteins by the immunogold method (for detailed description of the immunocytochemical methods, see [15]; Figures 2.5 and 2.12).

The light microscopy experiments were performed with primary human skin fibroblasts (HSF), L929 mouse fibroblasts and primary human keratinocytes (NHEK). The adhesion of the cells to the substrate was recorded by observing the distribution of the cells and the formation of focal contacts. Microstructured surfaces used had a groove width of 1, 2, 3, 5 and 10 μm and a depth of 0.4, 0.8, 1.6 and 2.0 μm. The keratinocytes did not reveal any orientation on the lattices tested (Figure 2.9). In addition, an increase of the number of focal contacts could not be observed on microstructured in comparison to smooth surfaces.

In contrast to human keratinocytes, human skin or gingiva fibroblasts did reveal a pronounced orientation ability as well as development of focal contacts depending on the texture of the microstructured substrate (Figure 2.10). However, this finding has been known for a long time [19–21].

Smooth substrate   Microstructured substrate

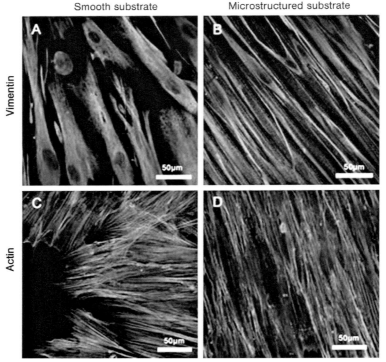

Figure 2.10 Fluorescence microscopy photographs of cells grown on microstructured surfaces. Antigens tested are indicated as well as the property of the microstructure. Above right: width of grooves 10 μm, depth of grooves 2 μm; below right: width of grooves 2 μm, depth of grooves 0.4 μm.

The description of the influence of geometrical properties of substrate lattices on the cellular behavior was not the only aim of the investigation. Moreover, we wanted to know whether or not "bioactive" coatings of implants would be able to alter cell biological parameters such as cell growth and migration. These coatings consist of components of the extracellular matrix (ECM) and provide a combination of factors which could be effective to modulate adhesion and growth (e.g. [22,23]).

Bioactive coatings (e.g. adhesion factors or growth factors) on implants are prone to abrasion due to shear stresses during implant insertion; for example, dental implants are usually inserted "press fit". In order to test whether surface structures are able to prevent abrasion, model implant surfaces modified by a combination of micro-structuring techniques and biological coatings (fibronectin $8\,\mu g\,cm^{-2}$) were subjected to shear stresses simulating conditions during implant insertion. Biological coatings were almost totally abraded on the ridges, while considerable amounts of the bioactive coatings remained in the grooves (Figure 2.11a).

The influence of mechanically stressed and unstressed fibronectin coatings on adhesion and spreading of cells was investigated and compared to the uncoated

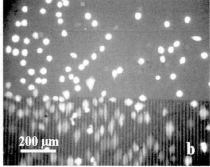

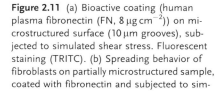

Figure 2.11 (a) Bioactive coating (human plasma fibronectin (FN, $8 \mu g \, cm^{-2}$)) on microstructured surface ($10 \mu m$ grooves), subjected to simulated shear stress. Fluorescent staining (TRITC). (b) Spreading behavior of fibroblasts on partially microstructured sample, coated with fibronectin and subjected to sim- ulated shear stress. Cells spread well on structured (lower part of sample) while cells on smooth, unstructured areas of the sample (upper half) showed a spherical appearance due to abrasion of the coating. Vital staining with FDA.

titanium surface. L 929 fibroblasts cultured on uncoated titanium surfaces showed mainly a spherical morphology. Biological functionalization of the surface by application of a fibronectin coating substantially improved adhesion and spreading of cells on smooth as well as structured areas of the samples. After application of mechanical shear stress the adhesion-promoting effect of the coating was maintained only on the structured part of the samples (lower half of Figure 2.11b) while cells on smooth, unstructured areas of the samples (upper half of Figure 2.11b) showed a rounded appearance indicating complete abrasion of the coating and its spreading– promoting effects. Control experiments revealed that these results were due to the coating and not caused by the surface structures [24,25].

The transmission electron microscopy investigation of the cells growing on microstructured lattices (Figure 2.5) was first of all dedicated to the description of the interaction between cell and substrate (Figure 2.12), including the detection of molecules involved in this interaction. For example, we investigated the expression and distribution of cytoskeleton-associated molecules and of molecules from which it is known that they contribute to the enlargement of the cellular surface. Enlargement of the cellular surface could be expected at least in such membrane domains directly attached to the microstructured lattice. This kind of investigation was performed at both the light and the electron microscopy levels.

For electron microscopy immunogold staining of intracytoplasmic antigens, a post-embedding procedure must be applied rather than the pre-embedding proce- dure which is only useful for the labeling of extracellular epitopes. However, in applying the post-embedding procedure, an artificial resin must be used which is different from Araldite, because Araldite polymerizes at high temperatures, which are detrimental for the maintenance of antigenicity. Performing post-embedding immunogold labeling, we used Lowicryl HM20 which polymerizes at minus degrees

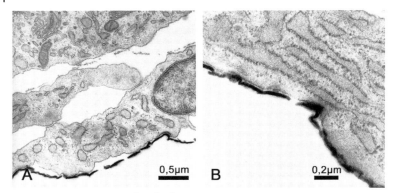

Figure 2.12 TEM micrographs of cells growing on microstructured silicon wafers replicated with Araldite and shadowed with titanium (compare Fig. 2.5). The titanium layer is seen as a broken black line which is closely associated to the surface of the cell.

under UV radiation after cryosubstitution of the cells (Figure 2.13). As already mentioned, we investigated not only cytoskeletal proteins such as vinculin but also proteins which are presumably involved in the adaptation of the membrane configuration under the constraints of geometrical (lattice-dependent) textures.

Members of the ERM protein family (ERM is an acronym meaning ezrin, radixin and moesin [26]) are frequently expressed where the cytoplasmic membrane is folded, which is the case particularly in lamellopodia of migrating cells or in microvilli of gastrointestinal epithelial cells [27]. We found that all three proteins were expressed by gingiva fibroblasts, but only one of them, radixin, showed an increased immunoreactivity against its antibody adjacent to the microstructure, whereas that against ezrin and moesin did not [16]. However, one has to be careful in

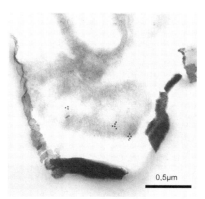

Figure 2.13 Gingiva-fibroblast, embedded in Lowicryl HM20, anti-vinculin immunogold labeling (15 nm gold grains). The labeling is particularly strong where the cell forms an enlargement of its surface according to the microstructure.

the interpretation of immunocytochemical preparations, because the proof of an increased expression on the protein or mRNA level would precede Western blotting or polymerase chain reaction (PCR), respectively.

On the other hand, during our studies we found reasons to doubt that the microstructure as such is responsible for the up-regulation of adhesion molecules: the material titanium itself seems to be causative for the induction of these molecules. Independent of whether titanium was smooth or microstructured, we observed a larger number of radixin-positive filopodia of cells than on glass or plastics (Figure 2.14).

Indeed, in cooperation with the group of Hans Schnittler in Dresden, Germany, we found that under shear stress conditions up to a shear force of 100 dyn cm^{-2} the cells adhered considerably better on titanium than on glass (Seebach, Schnittler, Pfeiffer and Wolburg, unpublished). However, these first experiments do not rule out the possibility that microstructured titanium surfaces may evoke an even better adhesiveness of the cells than a smooth titanium surface.

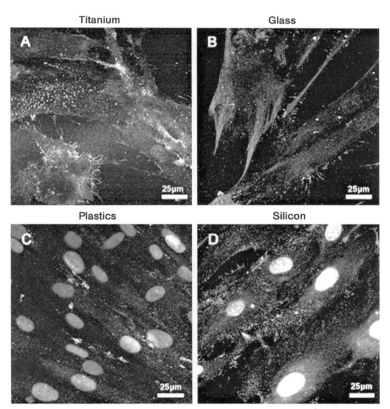

Figure 2.14 Immunoreactivity against radixin in human gingiva-fibroblasts, cultured on different materials without microstructures. On titanium, the immunolabeling was strongest.

Conclusions

1. The developed replica technique allows the transmission electron microscopy preparation of cells growing on titanium and the production of a large number of microstructured surfaces for the performance of cell culture experiments.
2. The adhesiveness and the dispersion of human fibroblasts, but not of keratinocytes, could be influenced by the microstructures. However, the material titanium seems to be even more important for adhesiveness than the microstructures themselves.
3. The combination of microstructures with different biological coatings (e.g. adhesion factors and growth factors) may be used to trigger specific cell responses in areas of the implant surface with different functionality.

2.2
Endothelial Cell-Specific Aspects in Biocompatibility Testing *In Vitro*

Kirsten Peters, Ronald E. Unger, Roman Tsaryk, Harald Schmidt,
Günter Kamp, and C. James Kirkpatrick

Vascular endothelial cells (EC) line the inner surface of all blood vessels. They form the interface between the circulating blood components and the surrounding tissues and are involved in the regulation of the blood–tissue barrier. EC are involved in inflammation by the expression of pro-inflammatory factors and EC are crucial for the process of blood vessel formation [28]. Thus, EC also have an influence on wound healing after implantation and long-term implant stability. Therefore, the maintenance of the full range of EC functions without influencing these functions inadvertently is a critical aspect in implant biocompatibility. The role of EC in inflammation can be described in the following way. After the initiation of inflammation by contact with inflammatory stimuli the release of pro-inflammatory factors by leukocytes (e.g. tumor necrosis factor-α (TNFα)) and EC (e.g. interleukin-8 (IL-8), monocytic chemoattractant protein-1 (MCP-1)) takes place. Because of this activation, circulating leukocytes come into loose contact with EC of the vessel wall (a process called rolling). Responsible for this are the so-called selectins (the endothelial selectin is called E-selectin). During the progression of inflammation the adhesion of leukocytes on the EC surface occurs. This adhesion is mediated by interactions of cell adhesion molecules on the leukocyte surface (the so-called integrins) and EC surface (e.g. intercellular adhesion molecule-1 (ICAM-1)) (Figure 2.15). Afterwards the transmigration of leukocytes into the surrounding tissues occurs. The interaction of different molecules on the EC and leukocyte surfaces (e.g. CD31) facilitates transmigration of leukocytes via the interendothelial contact [29].

Interendothelial contacts are involved in regulating barrier function. Different molecules such as occludin, ZO-1, VE-cadherin, CD31 and F-actin interact with each other to maintain this barrier. During inflammation, the interendothelial contacts are subjected to drastic changes, including the redistribution of CD31, VE-cadherin and the so-called peripheral actin ring. This redistribution leads to an increased permeability of the vessel wall, which facilitates transmigration of leukocytes across the endothelial lining [30].

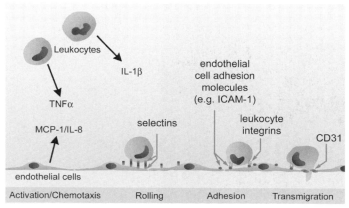

Figure 2.15 Schematic depiction of the different phases and factors in EC–leukocyte interaction during inflammation.

The vascular supply of tissues by angiogenesis is a crucial step during development. Moreover, angiogenesis is often a determining factor in pathophysiology (e.g. wound healing, tumor growth). Angiogenesis is regulated by complex interactions between EC and other cell types, and these interactions are mediated by the release of growth factors, molecules of the extracellular matrix (ECM), and cell surface adhesion molecules. An exemplary course of angiogenesis is shown in Figure 2.16: after initiation by pro-angiogenic factors (emanating from, for example, an area of inflammation or a growing tumor) proteolytic degradation of the surrounding ECM occurs. Moreover, the migration and proliferation of EC are necessary for the elongation of the growing blood vessel. Finally, recruitment of other cell types and the development of a new ECM are necessary for the stabilization of the new blood vessel [31].

Due to the role of EC in wound healing and to the requirements in biocompatibility testing we have developed different *in vitro* models for testing EC functionality, and examples of these EC-based *in vitro* models are elucidated within this section.

Materials and Methods
Human EC from juvenile foreskin were isolated and cultivated as described previously [32]. IL-8 and E-selectin were quantified as described before [33,34]. Fluorescent staining for ICAM-1, VE-cadherin, ZO-1, F-actin and nuclei was performed as described previously [35]. Cytotoxicity testing was carried out using the dye AlamarBlue™ (BioSource). Induction of angiogenesis *in vitro* was performed as described previously [36].

Results and Discussion: EC-Based *In Vitro* Models

EC-Derived Pro-Inflammatory Factors in Testing Biocompatibility *In Vitro*   Upon pro-inflammatory stimulation adhesion molecules are expressed on the EC surface and cytokines such as IL-8 or MCP-1 are released. These features contribute to the

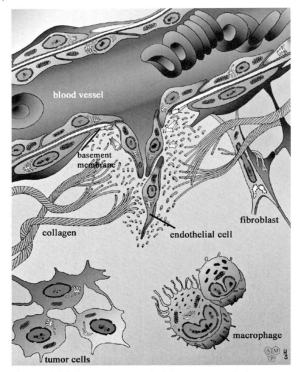

Figure 2.16 Schematic portrayal of the processes during angiogenesis. A growing tumor and a group of macrophages are depicted as exemplary inducers of angiogenesis in that they release a number of pro-angiogenic factors (with kind permission of AMP-Lab GmbH).

pro-inflammatory endothelial phenotype that permits the transmigration of leuko-cytes from the blood into the perivascular space [29].

This pro-inflammatory activation capability is maintained *in vitro* so that cultivated EC can be stimulated by pro-inflammatory factors (e.g. TNF-α) to express the above mentioned factors (ICAM-1 expression without (Figure 2.17a) and after (Figure 2.17b) TNF-α stimulation). An example for a quantification of immunocytochemi-cally stained cell cultures of such a pro-inflammatory factor is shown in Figure 2.18a. The exposure to Co^{2+} induced a concentration-dependent increase in E-selectin expression [37].

Furthermore, nanometer-scaled ceramic and metallic particles (nanoparticles) that induce pro-inflammatory activation in the lung of experimental animals [38] also displayed a pro-inflammatory effect in EC *in vitro*. An example for such a pro-inflammatory activation by nanoparticles is shown in Figure 2.18b. The exposure to polyvinylchloride (PVC) particles did not exhibit an effect on IL-8 expression, whereas all other particles tested (50 μg mL^{-1} of TiO_2, SiO_2, Co) induced an increase in IL-8 release. The degree of pro-inflammatory activation depended on the particle type [33].

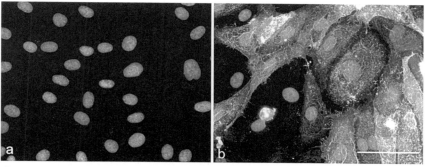

Figure 2.17 ICAM-1 expression in EC (a) without and (b) after TNF-α treatment (24 h, nuclei stained with Hoechst 33342, fluorescence microscopy, scale bar = 25 μm).

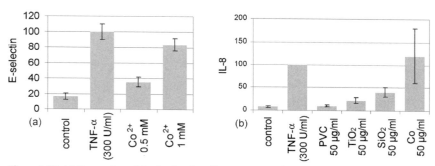

Figure 2.18 (a) Expression of E-selectin after 4 h treatment with TNF-α (positive control) and Co^{2+}; (b) IL-8 release after 24 h exposure to TNF-α and different nanoparticles (photometrical analysis: (a) enzyme immunoassay; (b) ELISA).

Thus, EC *in vitro* can be used to test the pro-inflammatory potential of biological factors and also for the evaluation of possible corrosion and wear products such as metal ions or particles [39].

Interendothelial Contacts EC have a number of characteristic molecules located within their intercellular contacts. Among them is CD31 that is distributed within the intercellular contacts as a continuous, often laminar and jagged band (Figure 2.19a). The distribution of VE-cadherin (Figure 2.19b), ZO-1 (Figure 2.19c) resembles that of CD31. Furthermore, the peripheral F-actin ring is located in close proximity to the interendothelial contacts (Figure 2.19d).

The pro-inflammatory stimulation of EC by, for example, TNF-α led to drastic changes of the interendothelial contact molecules: TNF-α stimulation led to the reduced and discontinuous distribution of CD31 (Figure 2.19e). Also the distribution of VE-cadherin, ZO-1 and F-actin was changed after TNF-α stimulation, leading to a massive reduction of VE-cadherin (Figure 2.19f) and ZO-1 (Figure 2.19g) and to cytoplasm-spanning F-actin fibrils which have been referred to as "stress fibers"

Untreated control   TNF-α   Co²⁺

CD31

VE-cadherin

ZO-1

F-actin

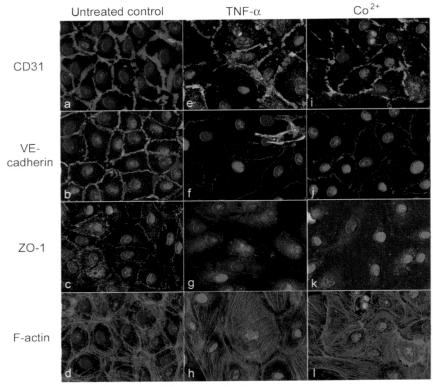

Figure 2.19 Pattern of interendothelial contact molecules (CD31, VE-cadherin, ZO-1 and F-actin) in non-treated EC *in vitro* ((a) CD31, (b) VE-cadherin, (c) ZO-1, (d) F-actin); in TNF-α-treated EC (24 h; (e) CD31, (f) VE-cadherin, (g) ZO-1, (h) F-actin); after exposure to 0.7 mM Co²⁺ ((i) CD31, (j) VE-cadherin, (k) ZO-1, (l) F-actin) (nuclei stained with Hoechst 33342, fluorescence microscopy, scale bar = 25 μm).

(Figure 2.19h). Thus, a pro-inflammatory stimulation of EC induced drastic changes in the distribution of intercellular contact molecules.

We utilized the analysis of the interendothelial contact pattern for testing the effects of toxic substances, e.g. Co²⁺. As shown by conventional cytotoxicity assays (alamarBlue™ and MTS test) the Co²⁺ concentration applied (0.7 mM) did not display a cytotoxic effect during 24 h of treatment [35]. However, the exposure of EC to 0.7 mM Co²⁺ induced striking changes of CD31 expression within the same time frame. The expression of CD31 (Figure 2.19i), VE-cadherin (Figure 2.19j) and ZO-1 (Figure 2.19k) within the intercellular contacts was drastically reduced and numerous F-actin stress fibers were formed (Figure 2.19l). Thus, the analysis of the interendothelial contact molecule pattern indicated an impairment of endothelial function with much higher sensitivity than conventional cytotoxicity assays, and the changes in the interendothelial phenotype reflect the effects of pro-inflammatory factors as well as toxic substances in a highly sensitive manner [35].

Model Systems of Angiogenesis *In Vitro* To elucidate the complexity of the angiogenic process *in vivo* a number of models to demonstrate angiogenesis *in vitro* have been developed. A major criterion for these *in vitro* models is the formation of capillary-like structures (CLS). Initially, the unambiguous quantification of the extent of angio-genesis remained a major problem. Manual counting of sprouts or the determination of sprout length was generally used for quantification of angiogenesis. Since these approaches were laborious they did not allow high throughput. To overcome these problems we have developed methods using software-supported analysis of micro-scopic images to determine the degree of angiogenesis *in vitro*. We have utilized different models of angiogenesis that mimic a number of the physiological aspects of angiogenesis *in vivo* and we have defined criteria for the software-supported recognition of angiogenic patterns [32,36].

Examples of materials- and corrosion product-induced effects on angiogenesis *in vitro* are shown in Figure 2.20. Using calcein-AM a contrast-rich staining of CLS over a dark background was achieved and thus it was evident that EC establish CLS *in vitro* after being embedded in a matrix consisting of fibrin and type I collagen (resembling the ECM in the implant periphery). The development of CLS was dependent on the stimulation by pro-angiogenic factors (VEGF/bFGF); without these factors most cells died (Figure 2.20a). In the presence of VEGF/bFGF an interconnected network of CLS evolved (Figure 2.20b). The deprivation of oxygen (called hypoxia) led to an increase of the amount and to an enlargement of vascular sprouts (Figure 2.20c) [36]. This is in agreement with the *in vivo* response in which the number of blood vessels increased in response to hypoxia [40]. In contrast, the presence of low amounts of Co^{2+} led to a downregulation of *in vitro* capillaries (Figure 2.20d). Interestingly, the contact to a Co28Cr6Mo alloy also impaired CLS formation (Figure 2.20e), whereas contact with cp-Ti did not have any deleterious effect (data not shown).

For the analysis of the microscopic images we utilized the software ImageTool 3.00 (IT). Image analysis functions of IT include dimensional (e.g. distance, perimeter, area) measurements. To choose the most applicable analysis parameter of angiogene-sis *in vitro*, we focused on the most conspicuous behavior of angiogenesis-stimulated EC *in vitro*. A major characteristic for the CLS formation was the aggrega-tion of single cells and morphogenesis to elongated objects, where most of the cells appeared optically fused. The utilization of these criteria demonstrated that the longer the interconnecting CLS were, the more pronounced was the degree of angiogenesis *in vitro*. From the analysis parameters included in IT the "object perimeter" was most useful for the quantification of these elongated, interconnected objects.

An example for object recognition with the software IT is shown in Figure 2.20f. The original image was analyzed by the "find object" command. All objects identified by the "find objects" procedure were further processed according to the attribute perimeter by the "object classification" plug-in. The classification procedure separated

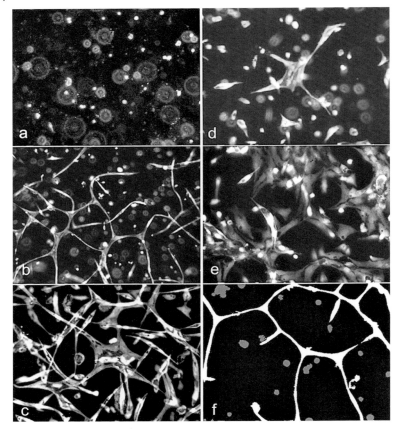

Figure 2.20 Angiogenesis *in vitro*: (a) without the addition of angiogenic factors; (b) bFGF/VEGF-stimulated; (c) bFGF/VEGF- and hypoxia-exposed; (d) bFGF/VEGF- and 0.1 mM Co^{2+}-exposed; (e) bFGF/VEGF-stimulated and grown in direct contact to a mirror-polished Co28Cr6Mo surface; (f) an example for object recognition with IT (after "find object" and "object classification" command, object with a perimeter above 250 pixels in white, objects with a perimeter below 250 pixels in dark grey).

objects with a perimeter above 250 pixels (here in white) from objects below this value (dark grey). Object classification resulted in the elimination of structures which could not be attributed to the angiogenic process (with perimeter <250 pixels).

The application of the software-supported image analysis with the chosen analysis parameter revealed the expected graded degree of angiogenesis *in vitro*: without pro-angiogenic factors a complete absence of CLS occurred (0% compared to the bFGF/VEGF-stimulated culture, set as 100%). The exposure to hypoxic conditions led to an increased angiogenic degree of 135%, whereas the exposure to 0.1 mM Co^{2+} induced a reduction of the angiogenic degree to approximately 50%. Thus, the software-supported image analysis offered the possibility of an unequivocal quantification of angiogenesis *in vitro* [36].

Conclusions

The assays demonstrated offer the possibility of testing biomaterial-induced effects on EC-specific features, such as the pro-inflammatory state and the ability to form blood vessel-like structures (pro-inflammatory and angiogenic phenotype respectively). We have shown that these *in vitro* systems are often more sensitive than a general measure of basic cellular functions. However, all *in vitro* models are merely an attempt to simplify the complex *in vivo* situation, and the models will inevitably show deviations from the *in vivo* situation. However, the *in vitro* systems described have demonstrated responses to factors *in vitro* similar to known responses *in vivo* and therefore should be useful in basic research and evaluation of biocompatibility.

Acknowledgments

This work was supported by the German Research Foundation (DFG, Priority Programme "Biosystem", KI 601/1-1–1-4). We would like to thank Susanne Barth for her excellent technical assistance and Bettina Hoffmann und Eva Eisenbarth for kindly supplying the metallic materials.

References

1 Monsees, T.K., Barth, K., Tippelt, S., Heidel, K., Gorbunov, A., Pompe, W. and Funk, R.H.W. (2005) Effects of different titanium alloys and nanosize surface patterning on adhesion, differentiation, and orientation of osteoblast-like cells. *Cells Tissues Organs*, **180**, 81–95.

2 Gerischer, H. (1989) Models for the discussion of the photo-electrochemical response of oxide layers on metals. *Corros. Sci.*, **29**, 257–266.

3 Roessler, S., Zimmermann, R., Scharnweber, D., Werner, C. and Worch, H. (2002) Characterization of oxide layers on Ti6Al4V and titanium by streaming potential and streaming current measurements. *Colloids Surf. B: Biointerfaces*, **26**, 387–395.

4 Mikhailov, A. and Gundersen, G.G. (1998) Relationship between microtubule dynamics and lamellipodium formation revealed by direct imaging of microtubules in cells treated with nocodazole or taxol. *Cell Motility Cytoskel.*, **41**, 325–340.

5 Ory, S., Destaing, O. and Jurdic, P. (2002) Microtubule dynamics differentially regulates Rho and Rac activity and triggers Rho-independent stress fiber formation in macrophage polykaryons. *Eur. J. Cell Biol.*, **81**, 351–362.

6 Funk, R.H.W. and Monsees, T.K. (2006) Effects of electromagnetic fields on cells: physiological and therapeutical approaches and molecular mechanisms of interaction. *Cells Tissues Organs*, **182**, 59–78.

7 Gorbunov, A.A., Pompe, W., Eichler, H., Huey, B. and Bonell, D.A. (1997) Nanostructuring of laser-deposited Ti films by self-limited oxidation. *J. Am. Ceram. Soc.*, **80**, 1663–1667.

8 Sickert, D. (2003) Biophotonische Anwendung Laser-strukturierter Titan-Titanoxid-Schichten Thesis, Institute of Material Sciences, University of Technology, Dresden.

9 Albrecht-Bühler, G. (1991) Surface extensions of 3T3 cells towards distant infrared light sources. *J. Cell. Biol.*, **114**, 493–502.

10 Bierbaum, S., Douglas, T., Hanke, T., Scharnweber, D., Tippelt, S., Monsees, T.K., Funk, R.H.W. and Worch, H. (2006) Collagenous matrix coatings on titanium implants modified with decorin and chondroitin sulfate: characterization and influence on osteoblastic cells. *J. Biomed. Mater. Res. A*, **77**, 551–562.

11 Erskine, L. and McCaig, C.D. (1997) Integrated interactions between chondroitin sulphate proteoglycans and weak dc electric fields regulate nerve growth cone guidance *in vitro*. *J. Cell Sci.*, **110**, 1957–1965.

12 Ohgaki, M., Kizuki, T., Katsura, M. and Yamashita, K. (2001) Manipulation of selective cell adhesion and growth by surface charges of electrically polarizes hydroxyapatite. *J. Biomed. Mater. Res. A*, **57**, 366–373.

13 Wilkinson, C.D.W., Curtis, A.S.G. and Crossan, J. (1998) Nanofabrication in cellular engineering. *J. Vac. Sci. Technol.*, **B 16**, 3132.

14 Pfeiffer, F., Kern, D., Scheideler, L., Geis-Gerstorfer, J. and Wolburg, H. (2002) Cell reactions to microstructured implant surfaces, *ICCE/9, 9th International Conference of Composites Engineering*, San Diego, CA, 615–616.

15 Pfeiffer, F., Herzog, B., Kern, D., Scheideler, L., Geis-Gerstorfer, J. and Wolburg, H. (2003) Cell reactions to microstructured implant surfaces. *Microelectron. Eng.*, **67/68**, 913–922.

16 Pfeiffer, F., Kern, D., Sotlar, K. and Wolburg, H. (2003) Expression of radixin and moesin in cells growing on artificial microstructured surfaces. *Biomaterialien*, **4**, 1393.

17 Augke, R., Eberhardt, W., Strähle, S., Prins, F.E. and Kern, D.P. (1999) Fabrication and characterization of Coulomb blockade devices in silicon. *Microelectron. Eng.*, **46**, 141.

18 Heeren, A., Burkhardt, C., Wolburg, H., Henschel, W., Nisch, W. and Kern, D. (2006) Preparation of nanostructured titanium surfaces for investigations of the interface between cell monolayers and titanium. *Microelectron. Eng.*, **83**, 1602–1604.

19 Meyle, J., Gültig, K., Hüttemann, W., Von Recum, A., Elssner, G., Wolburg, H. and Nisch, W. (1994) Oberflächen-mikromorphologie und Zellreaktion. *Z. Zahnärtzl. Implantol.*, **10**, 51–60.

20 Glass-Brudzinski, J., Perizzolo, D. and Brunette, D.M. (2002) Effects of substratum surface topography on the organization of cells and collagen fibers in collagen gel cultures. *J. Biomed. Mater. Res.*, **61**, 608–618.

21 Wieland, M., Chehroudi, B., Textor, M. and Brunette, D.M. (2002) Use of Ti-coated replicas to investigate the effects on fibroblast shape of surfaces with varying roughness and constant chemical composition. *J. Biomed. Mater. Res.*, **60**, 434–444.

22 Pettit, D.K., Hoffmann, A.S. and Horbett, T.A. (1994) Correlation between corneal epithelial cell outgrowth and monoclonal antibody binding to the cell binding domain of adsorbed fibronektin. *J. Biomed. Mater. Res.*, **28**, 685–691.

23 Brown, M.J. and Loew, L.M. (1996) Rho-dependent and -independent activation mechanisms of ezrin/radixin/moesin proteins: an essential role for poly-phoshoinositides *in vivo*. *Cell Motil. Cytoskel.*, **34**, 185–193.

24 Scheideler, L., Kern, D., Rupp, F., Weber, H. and Geis-Gerstorfer, J. (2002) Optimization of implant surfaces by a combination of microstructuring techniques and biological coatings, Poster presented at XVth Aachen Colloquium on Biomaterials, 27 Feb.–1 March.

25 Scheideler, L., Geis-Gerstorfer, J., Kern, D., Pfeiffer, F., Rupp, F., Weber, H. and Wolburg, H. (2003) Investigation of cell reactions to microstructured implant surfaces. *Mater. Sci. Eng. C*, **23**, 455–459.

26 Yonemura, S., Matsui, T., Tsukita, S. and Tsukita, S.J. (2002) Rho-dependent and -independent activation mechanisms of ezrin/radixin/moesin proteins: an essential role for polyphoshoinositides *in vivo. J. Cell Sci.*, **115**, 2569–2580.

27 Bretscher, A., Edwards, K. and Fehon, R.G. (2002) ERM proteins and merlin: integrators of the cell cortex. *Nature Rev. Mol. Cell Biol.*, **3**, 586–599.

28 Feletou, M. and Vanhoutte, P.M. (2006) Endothelial dysfunction: a multifaceted disorder (The Wiggers Award Lecture) *Am. J. Physiol. Heart Circ. Physiol.*, **3**, H985–H1002.

29 Peters, K., Unger, R.E., Brunner, J. and Kirkpatrick, C.J. (2003) Molecular basis of endothelial dysfunction in sepsis. *Cardiovasc. Res.*, **60** (1), 49–57.

30 Simionescu, M. and Antohe, F. (2006) Functional ultrastructure of the vascular endothelium: changes in various pathologies. *Handb. Exp. Pharmacol.*, **176**, 41–69.

31 Battegay, E.J. (1995) Angiogenesis: mechanistic insights, neovascular diseases, and therapeutic prospects. *J. Mol. Med.*, **73** (7), 333–346.

32 Peters, K., Schmidt, H., Unger, R.E., Otto, M., Kamp, G. and Kirkpatrick, C.J. (2002) Software-supported image quantification of angiogenesis in an *in vitro* culture system: application to studies of biocompatibility. *Biomaterials*, **23** (16), 3413–3419.

33 Peters, K., Unger, R.E., Gatti, A.M., Monari, E. and Kirkpatrick, C.J. (2004) Effects of nano-scaled particles on endothelial cell function *in vitro*: studies on viability, proliferation and inflammation. *J. Mater. Sci. Mater. Med.*, **15** (4), 321–325.

34 Peters, K., Unger, R.E., Gatti, A.M., Sabbioni, E., Gambarelli, A. and Kirkpatrick, C.J. (2006) Nanomaterials: Toxicity, Health and Environmental Issues, VCH, Weinheim, pp. 198.

35 Peters, K., Unger, R.E., Stumpf, S., Schaefer, J., Tsaryk, R., Hoffmann, B., Eisenbarth, E., Breme, J., Ziegler, G. and Kirkpatrick, C.J. , Cell-type specific aspects in biocompatibility testing: the intercellular contact *in vitro* as an indicator for endothelial cell compatibility, *J. Mater. Sci. Mater. Med.* (in press).

36 Peters, K., Schmidt, H., Kamp, G., Unger, R.E., Hoffmann, B., Stenzel, F., Ziegler, G., Jung, M., Stumpf, S. and Kirkpatrick, C.J. (2005) Software-Supported Quantification of Angiogenesis in an *in vitro Culture System*. *Examples of Application in Studies of Basic Research, Biocompatibility and Drug Discovery*, Nova Science Publishers, 103.

37 Kirkpatrick, C.J., Barth, S., Gerdes, T., Krump-Konvalinkova, V. and Peters, K. (2002) Pathomechanisms of impaired wound healing by metallic corrosion products. *Mund Kiefer Gesichtschir.*, **6** (3), 183–190.

38 Dick, C.A., Brown, D.M., Donaldson, K. and Stone, V. (2003) The role of free radicals in the toxic and inflammatory effects of four different ultrafine particle types. *Inhal. Toxicol.*, **15** (1), 39–52.

39 Kirkpatrick, C.J., Unger, R.E., Krump-Konvalinkova, V., Schmidt, H., Kamp, G. and Peters, K. (2003) Experimental approaches to study vascularization in tissue engineering and biomaterial application. *J. Mater. Sci. Mater. Med.*, **14** (8), 677–681.

40 Howell, K., Preston, R.J. and McLoughlin, P. (2003) Chronic hypoxia causes angiogenesis in addition to remodelling in the adult rat pulmonary circulation. *J. Physiol.*, **547** (1), 133–145.

3
Topology-Dependent Cellular Interactions

3.1
Cell Structure–Cell Function Dependence on Titanium Surfaces

J.G. Barbara Nebe, Frank Lüthen, and Joachim Rychly

The interaction of cells with implant materials at the interface is decisive for the clinical success of an implant. Because of the progress made in cell biology, biocompatibility investigations are increasingly focused on primary cellular mechanisms which control cell physiology. Processes that play a role when cells establish themselves on a material surface involve initial attachment and adhesion [1] followed by cell spreading and migration. This first phase of cell–material interaction will influence the cell's capacity to proliferate and differentiate [2]. The molecules responsible for these processes involve adhesion receptors such as integrins, the actin cytoskeleton and cytoskeletally associated proteins like vinculin [3,4].

Our investigations were focused on the question as to whether modifications of the surface topography of pure titanium can influence cellular adhesion structures, which in consequence could lead to alterations in cellular functions.

Materials and Methods
Titanium of technical purity (grade 2) was used as discs of 30 mm in diameter. The physical structure of the surface was modified by polishing (P) with SiC wet grinding paper (grit P4000), machining (non-treated) (NT) and blasting with corundum (aluminum oxide) particles (CB) (500–600 µm) at 6 bar [5–7]. Roughness measurements – average roughness (R_a, ISO 4287/1) – were performed with the surface profiler HOMMEL-Taster T8000 (Hommel) [5].

Human Osteoblasts Human primary osteoblasts (pOb) were derived from trabecular bone of the femoral head of orthopedic patients [6,8]. The pOb cells as well as the MG-63 osteoblastic cells (ATCC, LGC Promochem) were cultured in DMEM with 10% fetal calf serum and 1% gentamicin (Ratiopharm GmbH) at 37 °C and in a 5% CO_2 atmosphere. Titanium discs were placed into 6-well plates (Greiner Bio One). As control, collagen I (COL, rat tail, 20 µg cm^{-2}) (TEBU) coated cover glass was used. In general, cells were seeded with a density of 3×10^4 cells cm^{-2} on the specimens.

Metallic Biomaterial Interfaces. Edited by J. Breme, C. J. Kirkpatrick, and R. Thull
Copyright © 2008 WILEY-VCH Verlag GmbH & Co. KGaA, Weinheim
ISBN: 978-3-527-31860-5

Cell Spreading Osteoblasts were cultured for 24 h, trypsinized and the cell membrane was stained with the red fluorescent linker PKH26 (PKH26 General Cell Linker Kit, Sigma) in suspension. The cells were then cultured for 3, 16, 24, and 40 h on the structured titanium discs [6,9].

Time-Dependent Organization of the Actin Cytoskeleton MG-63 cells were cultured on the titanium specimens for 20, 40, 60 and 120 min, fixed with 4% PFA, permeabilized with 0.1% TritonX-100 (Merck), stained with phalloidin-TRITC (diluted 1:100, Sigma) and analyzed with an LSM 410.

Colocalization of Integrin Receptors and Actin POb cells were cultured for 48 h on titanium plates, washed with PBS containing Ca^{2+} and Mg^{2+}, incubated with mouse anti-human-β1 (CD29, K20 clone, diluted 1:6, Immunotech) and labeled with AlexaFluor® 488 secondary goat anti-mouse IgG (1:300, Molecular Probes) [6,9]. Then cells were fixed and permeabilized with ice-cold acetone followed by actin staining with phalloidin-TRITC (diluted 1:100, Sigma). Cells were analyzed with an LSM Leica TCS SP2 AOBS (Leica Microsysteme).

Colocalization of Vinculin and Actin Osteoblasts were cultured for 48 h, fixed with 4% PFA and permeabilized with 0.1% Triton-X100. Cell staining was according to [1,10]: Briefly, cells were incubated with the primary antibody mouse-anti-vinculin (1:100, Sigma) followed by incubation with the secondary Cy3-tagged anti-mouse antibody (1:200, Jackson Immuno Research Lab). The actin cytoskeleton was stained with BODIPY FL phallacidin (1:40, Molecular Probes) and analyzed with an LSM Leica.

Double Immunofluorescence of Fibronectin and Actin MG-63 cells were cultured for 48 h on the specimens, fixed with 4% PFA (10 min) and incubated with a primary rabbit anti-fibronectin antibody (1:40, Sigma), followed by the secondary Cy3-tagged anti-rabbit antibody (1:200, Jackson Immuno Research Lab). For actin staining cells were permeabilized with TritonX-100 and incubated with BODIPY FL phallacidin (1:40, Molecular Probes) [6].

Mineralization The *in vitro* mineralization was quantitatively analyzed using calcein [11] which binds to calcium phosphate synthesized by osteoblasts. MG-63 cells were cultured for 21 d in DMEM supplemented with 10 mM β-glycerophosphate and 50 µM mL^{-1} ascorbic acid on the titanium samples in 24-well plates. Cells were cultivated for additionally 24 h with 5 µl mL^{-1} calcein (Sigma). For control experiments cells were cultured in parallel without calcein. The calcein fluorescence was measured in a fluorescence reader (CytoFluor™ 2350, Millipore) using an excitation and emission wavelengths of 485 and 530 nm, respectively. Because of the autofluorescence of the cell monolayer the fluorescence values of samples without calcein were subtracted from fluorescence values of calcein-incubated samples [12]. These values of NT and CB were then related to P (=100%) to obtain the percentage increase/decrease of mineralization due to the topography.

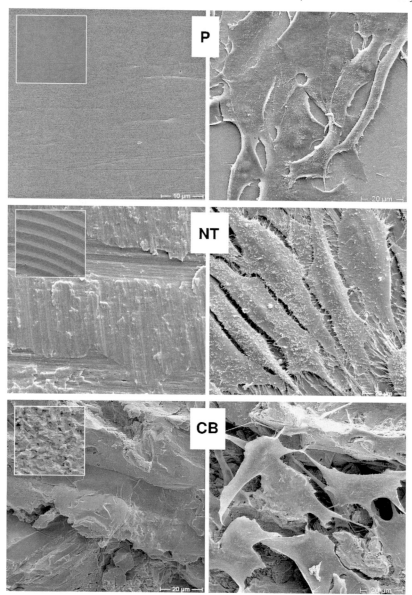

Figure 3.1 Left: SEM images of pure titanium with different surface roughnesses: polished (P) $R_a = 0.19\,\mu m$; machined non-treated (NT) $R_a = 0.54\,\mu m$ and corundum-blasted (CB) $R_a = 6.07\,\mu m$. Note that the surface of NT reveals grooves and striations related to machining. The surface of CB consists of sharp edges and ridges. (Insets document a magnification of ×100). Right: SEM images of MG-63 cells on these structured titanium surfaces. Note that on NT the cell growth is directed due to the striations, and on CB titanium the osteoblastic cells span through the ridges and MG-63 cells reveal a cuboidal shape with long cellular extensions.

Results

The pure titanium surfaces with modified topographies revealed the following R_a values: 0.19, 0.54, and 6.07 μm for polished (P), machined (NT) and corundum-blasted (CB) titanium, respectively [5,6]. The SEM images (Figure 3.1) indicated that the titanium surface of P is smooth resulting in a flattened morphology of MG-63 osteoblasts. The surface of NT featured grooves and striations related to machining, which resulted in a directed cell growth. The CB titanium surface showed sharp ridges and edges where the cells spanned through and revealed a cuboidal shape with long cellular extensions.

Cell spreading is a time-dependent process. MG-63 cells were significantly impaired in their cell areas on the roughest titanium CB (Figure 3.2) within 40 h compared to the more smooth titanium surfaces ($p < 0.01$). The development of cellular structures depending on the time is also determined by the surface topography, which is seen for actin within 120 min in Figure 3.3: whereas on P and NT the actin was pronounced in well-defined stress fibers comparable to a COL-coated surface, osteoblasts on the sharp-edged CB revealed not only a decelerated actin filament organization but also an irregular distribution which appears fragmented. The colocalization was influenced by the topography of the sharp-edged titanium CB: the integrin adhesion receptors as well as the adapter protein vinculin were not colocalized with the actin filaments in pOb osteoblasts (Figure 3.4).

In this context we are interested in whether alterations in cellular adhesion structures due to the titanium topography were the cause of alterations in specific cellular functions like fibronectin formation or mineralization. The double immuno-fluorescence images in Figure 3.5 revealed that the organization of FN even on the

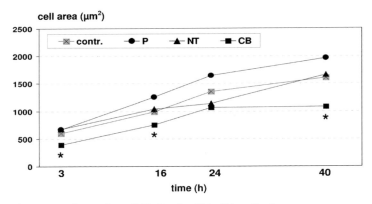

Figure 3.2 Cell spreading of MG-63 cells within 40 h on titanium with different surface topography compared to the collagen control. Note that spreading of MG-63 cells on CB is significantly impaired at 3, 16 and 40 h (* $p < 0.01$; unpaired samples *t*-test, $n = 40$) (LSM 410).

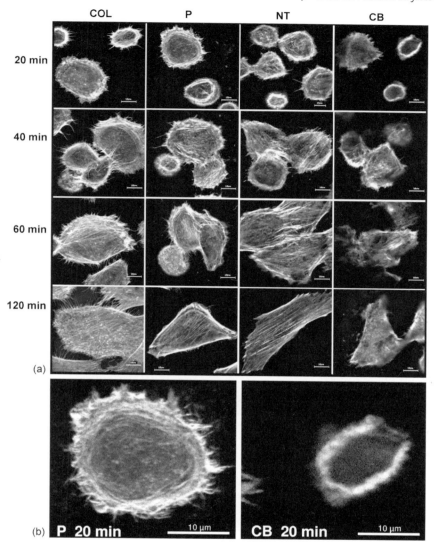

Figure 3.3 (a) Time-dependent development of actin filaments in MG-63 cells on titanium with different surface topography. Note that the development of actin filaments on the more smooth surfaces P and NT is enhanced compared to the sharp edged CB titanium. (b) Already after 20 min the actin on P demonstrates small stress fibers whereas on CB the actin is only submembraneously organized.

Integrin and Actin

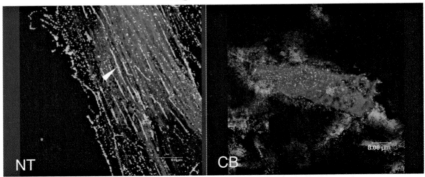

Vinculin and Actin

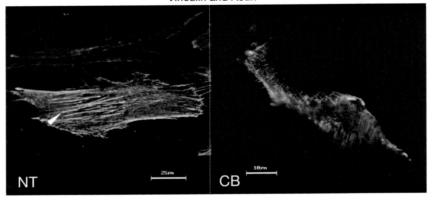

Figure 3.4 Colocalization of the actin cytoskeleton with adhesion components of primary osteoblasts on titanium with different surface topography. Top: β1-integrins (green) and actin (red). Note that the colocalization on CB is impaired and, furthermore, the integrins on CB are in punctiform distribution and the actin filaments are irregularly distributed. On titanium surfaces with lower R_a values, as is shown here for NT, the fibrillar integrin adhesions are strongly colocalized at sides of actin fibers (arrowhead) which are directed according to the grooves of the machined surface. (Confocal microscopy, LSM Leica.) Bottom: Adapter protein vinculin (red) and actin (green). Note that the colocalization on CB with its sharp edges is impaired, whereas on NT vinculin is well organized and colocalized at the tips of the actin filaments (arrowhead) and directed due to the grooves of the surface.

apical side of the cell surface demonstrated similarities to the alignment of the actin filaments. Especially on NT, fibronectin formed fibrils that were parallel to the direction of the actin cytoskeleton. On P, cells also formed FN fibrils both on the basal side (not shown) and apically; however, these fibrils were not directed because the material surface was smooth without striations and therefore the material did not determine the direction of actin filaments. On the roughest surface CB the FN formation on the apical side appeared more punctiform and was clustered, correlating with the irregular

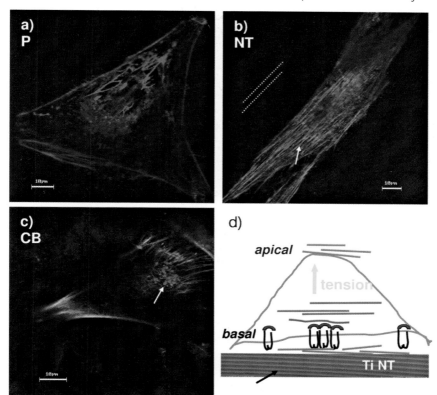

Figure 3.5 (a–c) Double immunofluorescence of fibronectin (red) and actin (green) on the apical side of MG-63 osteoblastic cells on titanium with different surface topography. Note that on NT titanium (b) FN forms fibrils that are directed in parallel to the actin cytoskeleton (arrow). Lines (grey) indicate the direction of material striations due to machining. On CB (c) the FN formation apically appears more punctiform and is clustered correlating with the irregular distribution of the actin. (Confocal microscopy, LSM 410.) (d) The scheme represents a cell on titanium with directed grooves (arrow) (e.g. NT). This topographical feature determines the fibrillar adhesions of integrins (black) (see also Fig. 3.4) and the aligned actin cytoskeleton (green) which results in basally and apically aligned FN fibrils (red) due to the tension throughout the cell.

distribution of actin. The analysis of mineralization capacity (Figure 3.6) of MG-63 osteoblasts also revealed that cells were functionally impaired when cell structures were altered: on titanium CB the mineralization was significantly decreased ($p < 0.005$).

Discussion

Because integrin receptors physically interact with the actin cytoskeleton [13] and stress fibers of actin function as transmitter of forces [14] as well as signal transducers [4], thus responsible for the cell function, we investigated the influence of titanium surfaces with different topographies on the formation of integrin adhesions

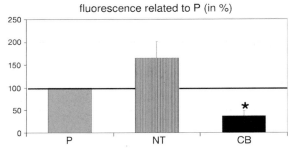

Figure 3.6 Mineralization of MG-63 osteoblasts on titanium with different surface topographies. Note that mineralization of osteoblasts grown on corundum-blasted titanium is significantly reduced (paired *t*-test CB:NT, $^*p < 0.005$, mean ± SEM, $n = 8$) which is possibly due to the impaired adhesion structures of the cells on this topography. (Fluorescence reader CytoFluor.)

and actin cytoskeleton in human osteoblasts. A pronounced alignment of actin filaments occurred on NT titanium with its directed grooves due to machining. Thus, the material surface of NT determined not only the direction of cellular growth and morphology as shown by SEM images, but also the alignment of actin throughout the whole cell. In contrast, on CB titanium with its sharp ridges actin filament formation was influenced and actin appeared irregularly distributed. In addition, the colocalization of integrins as well as vinculin with the actin cytoskeleton was impaired. These differences in cellular structures in response to the surface roughness of titanium led to consequences for FN fibrillogenesis: on the NT surfaces FN was formed in fibrillar structures which were directed parallel to the alignment of the actin cytoskeleton. Fibrils of FN were bound to integrins in parallel with actin bundles [15]. We observed the strongly aligned fibrils of FN not only on the basal side but interestingly also on the apical surface of a cell, indicating a cellular tension throughout the cell. Interesting investigations concerning cellular tension revealed that different integrins – among them β1 – can mediate force transfer to the cytoskeleton, and that this cytoskeletal tension was transferred to the extracellular matrix [16,17]. Other investigations suggest that the topography is dominant over the chemistry of a material surface [18] and demonstrated a sensitivity of osteoblasts due to the roughness of titanium surfaces [19–21]. Our results evidence that the impairment of specific osteoblastic functions, as was seen for mineralization of osteoblasts on the rough, sharp-edged CB titanium surface, may be caused by alterations in the adhesion structures of these cells due to the influence of the topography.

Acknowledgments
The authors are very thankful to the DOT GmbH Rostock for providing us with structured titanium. We acknowledge the technical help of Gerhard Fulda of the Electron Microscopic Centre of the Medical Faculty of the University of Rostock.

3.2
Cell Behavior on Micro- and Nanostructured Surfaces

3.2.1
Cell Biological Response on Polished, Structured and Functionally Coated Titanium Alloys

Eva Eisenbarth, Margit Müller, Stefan Winter, Dirk Velten, and Frank Aubertin

Biocompatibility tests (cell viability and cell function tests) were carried out with MC3T3-E1 (murine calvaria osteoblast-like cells), GM7373 (bovine aortic endothelial cells) and L132 (human embryonic lung epithelial cells) which were seeded on 15 mm diameter slices in 12-well plates and were cultured for seven days on the material samples. Cell function tests (spreading, collagen I production, cell migration and actin cytoskeleton staining) were performed on the materials described in Section 2.1 of Part I. For the migration tests, the cell monolayer (which was formed on the samples after seven days) was detached from cells within a gap of 0.5 mm width by means of a cell scraper. After 24, 36 and 48 h the migration of the cells from the bordering monolayer in the cell-free gap was observed. The spreading of the cells was analyzed after staining the cells on the samples with May–Gruenwald solution. The collagen I synthesis of the osteoblasts on the samples was compared by coloring the collagen I with Sirius Red stain.

Alloys
Cell viability tests were carried out using the near-β-alloys Ti13Nb13Zr, Ti30Nb and Ti30Ta and the β-alloy Ti15Mo5Zr3Al (Section 2.1 of Part I) and on cp-Ti and

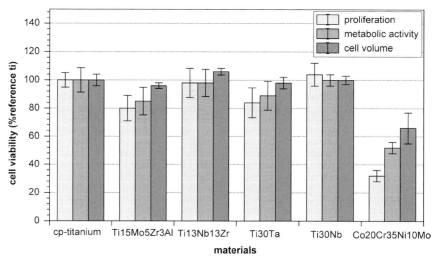

Figure 3.7 Viability of MC3T3-E1 osteoblasts on cp-Ti and Ti13Nb13Zr.

Co20Cr35Ni10Mo for comparison (Figure 3.7). The proliferation and the metabolic activity on Ti13Nb13Zr and Ti30Nb are comparable to that of uncoated cp-Ti, but are slightly decreased on Ti30Ta and Ti15Mo5Zr3Al. The implant material Co20Cr35-Ni10Mo exhibits the poorest results concerning both proliferation rate and metabolic activity.

Oxide Layers

Cell viability tests were also performed on anodically and thermally oxidized samples of the near-β-alloy Ti13Nb13Zr and the β-alloy Ti15Mo5Zr3Al and on Co20Cr35-Ni10Mo for comparison (Figure 3.8).

Subsequent to the thermal and anodic oxidation distinct differences in the biocompatibility of the β- and near-β-titanium alloys can be observed. Particularly the cytocompatibility of the alloy Ti15Mo5Zr3Al is clearly reduced as compared to that of cp-Ti, whereas the biocompatibility of the alloy Ti13Nb13Zr is only slightly decreased. Co20Cr35Ni10Mo shows the poorest results with regard to proliferation rate and metabolic activity.

Investigations for the determination of the inflammatory potential were carried out on anodically and thermally oxidized samples of cp-Ti and on glass substrates as reference materials by culturing of microvascular aortic endothelial cells (Figure 3.9).

In inactive nonstimulated endothelial cells, the actin distribution is in accordance with the allocation of CD31 (an integrin of endothelial cell contacts). This connection of actin and CD31 as a marker of a non-proinflammatory status is seen in endothelial cells grown on the glass substrate. On polished titanium, the connection between

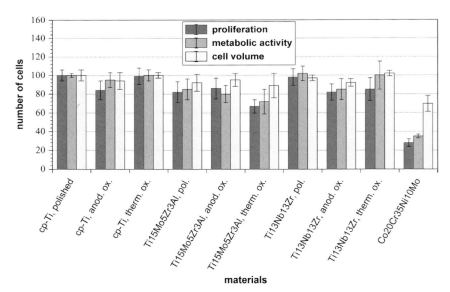

Figure 3.8 Viability of MC3T3-E1 osteoblasts on polished and thermally and anodically oxidized cp-Ti, Ti13Nb13Zr, Ti15Mo5Zr3Al and on Co35Ni20Cr10Mo for comparison.

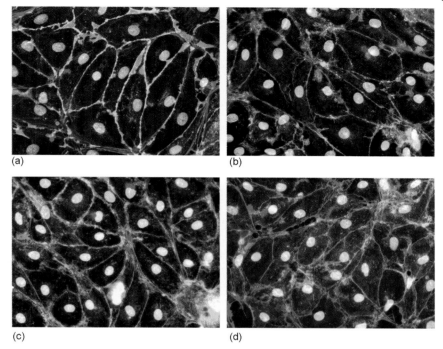

(a)   (b)

(c)   (d)

Figure 3.9 Actin (red) and CD31 (green) distribution of microvascular aortic endothelial cells cultured on (a) glass, (b) polished cp-Ti, (c) anodically oxidized cp-Ti and (d) thermally oxidized cp-Ti.

both molecules is not yet tight. The CD31 are partly disconnected from the actin fibers at the cell fringe. The thermally oxidized titanium samples show hardly any CD31 at the cell fringe and they shift to the cell surface. This is a distinct indication of a proinflammatory status and an activation of the endothelial cells adhering to the thermally oxidized titanium surface.

Cell viability tests were also carried out on sol–gel coatings (see Section 2.2 of Part I). Figure 3.10 shows the morphology. The appearance of cells on TiO_2- and Nb_2O_5-coated samples is good and comparable to that on cp-Ti. The cells on Ta_2O_5 and ZrO_2 are also viable and in a good condition; the spreading however is slightly decreased which can indicate a decreased adhesion. The results of the viability tests are summarized in Figure 3.11.

The quantitative biocompatibility tests confirm the morphology results. TiO_2 and Nb_2O_5 show a good, and to cp-Ti comparable, biocompatibility, whereas Ta_2O_5 and ZrO_2 exhibit a slightly decreased vitality.

These results led to further investigations on the biocompatibility of the most promising sol–gel layers TiO_2 and Nb_2O_5. With these layers migration (Figure 3.12) and spreading tests (Figure 3.13) were also carried out.

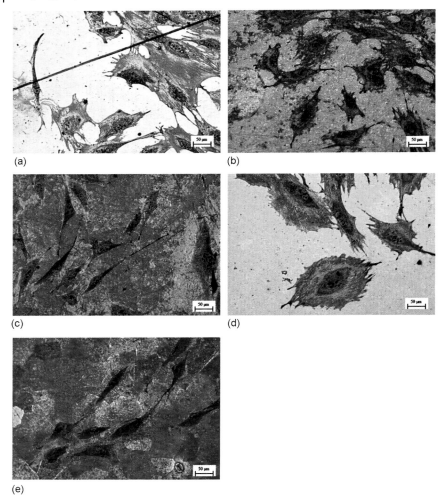

Figure 3.10 Morphology of MC3T3-E1 osteoblast-like cells on sol–gel oxide layers: (a) cp-Ti, (b) Nb_2O_5, (c) Ta_2O_5, (d) TiO_2, (e) ZrO_2.

The investigation of cell migration, spreading and collagen I production reveals a lower migration rate on the Nb_2O_5-coated samples but an increased spreading of the osteoblast-like cells on the niobium oxide coatings.

Additional tests showed that the collagen I production of the osteoblasts is slightly increased on TiO_2 and greatly increased on Nb_2O_5 as compared to uncoated cp-Ti. The cells require a longer time to cover a gap in a monolayer in contact with niobium oxide as compared to the cell migration on uncoated cp-Ti and TiO_2-coated slices. However the MC3T3-E1 cells show better spreading and an increased collagen I production on the niobium oxide coatings. This acts as a precursor of bone remodeling *in vivo*.

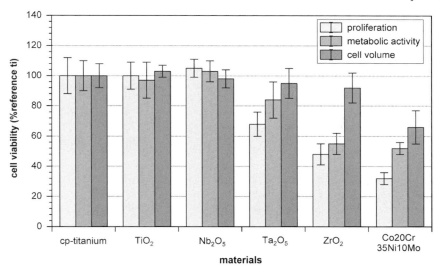

Figure 3.11 Viability tests of osteoblast-like cells on various sol–gel oxide layers on cp-Ti.

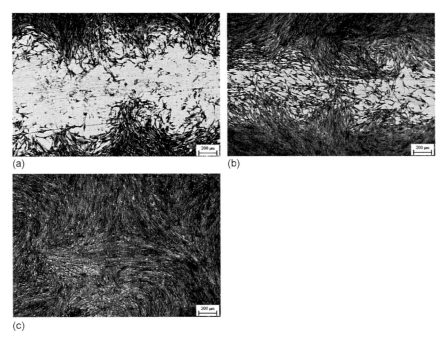

Figure 3.12 MC3T3-E1 migration on (a) cp-Ti, (b) Nb_2O_5, (c) TiO_2.

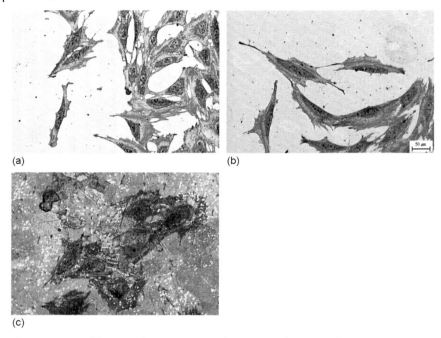

(a)   (b)

(c)

Figure 3.13 Osteoblast spreading on (a) cp-Ti, (b) TiO$_2$, (c) Nb$_2$O$_5$-coated cp-Ti.

Nanostructured Oxide Layers

In a first step the adhesion kinetics of MC3T3-E1 osteoblast-like cells was determined on variously nanostructured Nb$_2$O$_5$ surface layers. The roughnesses were 7, 13 and 39 nm as described in Section 2.2 of Part I. On these samples and on polystyrene as a reference material approximately 15 000 osteoblast-like cells were cultured. After 15,

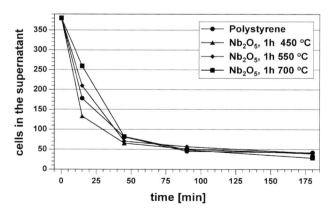

Figure 3.14 Cell number in the supernatant of Nb$_2$O$_5$-coated cp-Ti samples and of polystyrene as a reference material.

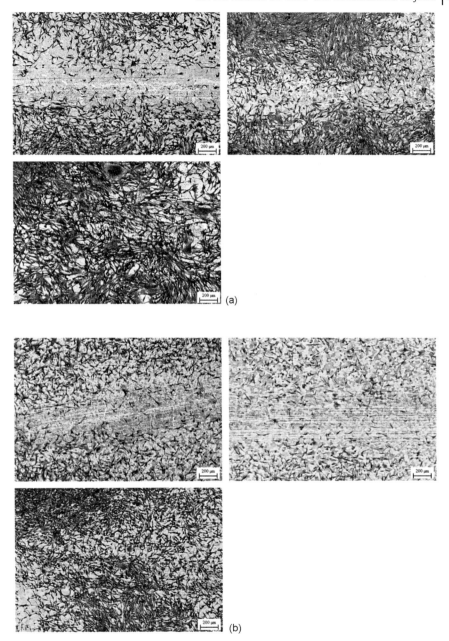

Figure 3.15 Migration of osteoblast-like cells on variously structured Nb_2O_5 layers.

Figure 3.15 (continued)

45, 90 and 180 min the remaining number of cells in the supernatant was determined (Figure 3.14); the others had adhered on the surface.

After 15 min in the state of primary adhesion, the majority of the cells adhered on Nb_2O_5 (1 h, 450 °C, $R_a = 7$ nm) followed by polystyrene, Nb_2O_5 (1 h, 500 °C, $R_a = 13$ nm) and Nb_2O_5 (1 h, 700 °C, $R_a = 39$ nm). After 45, 90 and 180 min an influence of the surface structure on the adhesion of the cells can no longer be detected. The number of cells attached to the surfaces was almost identical. Figure 3.15 shows the results of the migration tests.

Differences in the resettlement of the gap can be seen after 36 h. On Nb_2O_5 layers (450 °C, $R_a = 7$ nm) first cell bridges could be observed, whereas on the other samples only single cells had appeared in the gap. After 48 h the gap on the Nb_2O_5 layers (450 °C) is almost completely closed, whereas on the more strongly closed structured surfaces a thin mesh-like cell formation had taken place. An explanation for this behavior could be the good adhesion of the osteoblast-like cells on rough substrates and therefore a good mobility on smoother surfaces.

Actin staining was carried out on the variously structured Nb_2O_5 layers after culturing of osteoblast-like cells for two days.

Due to the appearance of the actin filament (thickness and number of fibers (stress fibrils) and their organization (distribution of the fibers in the direction of the cell spreading)) conclusions on the cell adhesion can be drawn, because the actin fibers build up the connections of the cells with the extracellular matrix and the substrate. Thus the investigation of the actin skeleton is an established method for indirect cell adhesion tests.

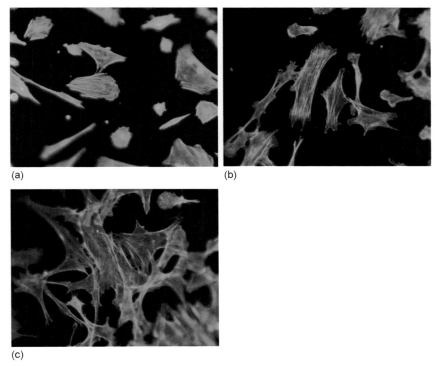

(a)

(b)

(c)

Figure 3.16 Actin staining of osteoblast-like cells on differently structured Nb$_2$O$_5$ layers on cp-Ti: (a) 450 °C, $R_a = 7$ nm, (b) 550 °C, $R_a = 13$ nm, (c) 700 °C, $R_a = 39$ nm.

Micrographs of the stained cells are shown in Figure 3.16. The optimum formation of the actin cytoskeleton of the cells is found on Nb$_2$O$_5$ (550 °C). The actin fibers are organized in stress fibrils and the cells show a good interaction with their neighbors, which can be seen by the large amount of focal contacts. The osteoblasts are well spread and adhere strongly to the substrate. The actin fibers on Nb$_2$O$_5$ (700 °C) are less organized than on Nb$_2$O$_5$ (550 °C) and exhibit a minor organization as stress fibrils, which can indicate a weaker adhesion to the substrate. The interaction with adjacent cells is comparable to that of Nb$_2$O$_5$ (550 °C).

The poorest production of actin fibers is seen on the amorphous Nb$_2$O$_5$ (450 °C) layers. On these substrates fewer connections of the cytoskeleton to the focal contacts produced by the actin fibers are observed. Thus, a weaker adhesion of the cells than in the case of the crystalline substrates can be expected.

Microstructured, Hemocompatible Surfaces
A hemocompatible stent material is intended to suppress the neointimal proliferation of the surface of an implant material but must not act cytotoxically. In order to investigate the influence of the various functional coatings (see Chapter 3 of Part I)

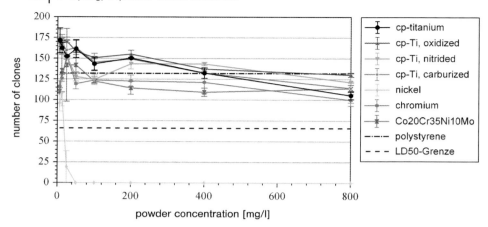

Figure 3.17 LD_{50} test with L132 fibroblasts on differently coated titanium powders compared with powders of chromium, nickel and Co20Cr35Ni10Mo.

on their cytotoxicity LD_{50} tests were carried out. In these tests L132 human lung epithelial cells were brought into contact (7 days) with a suspension (6, 12, 25, 50, 100, 200, 400 and 800 mg L^{-1}) of titanium powders (20–50 μm in diameter) which had been oxidized, nitrided and carburized. As reference materials powders of cp-Ti, chromium, nickel and the alloy Co20Cr35Ni10Mo were tested. After the testing the colonies were dyed in May–Gruenwald solution and counted. The results are shown in Figure 3.17. It is shown that all titanium based materials possess a good biocompatibility, whereas the negative reference material nickel reveals explicit cytotoxic properties.

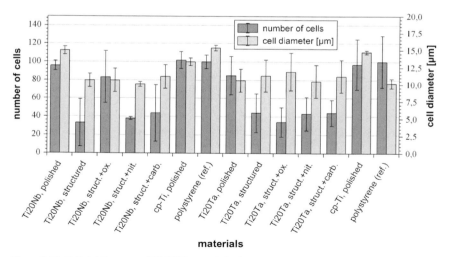

Figure 3.18 Cell viability tests of GM7373 on polished, structured and (structured + coated) Ti20Nb and Ti20Ta and on polished cp-Ti and polystyrene as reference materials.

Cell viability tests concerning proliferation rate and cell volume were carried out with GM7373 bovine aortic endothelial cells. These had been cultured at 37 °C on differently structured and coated substrates and characterized after 5 days as to proliferation rate and cell volume. The results for Ti20Nb and of Ti20Ta samples which are representative of all materials tested are shown in Figure 3.18.

The results show that the proliferation rate and the cell diameter on the polished samples correspond to those on the reference polystyrene. By structuring and coating the proliferation rate of endothelial cells was drastically reduced without reducing the cell volume which can be an indicator for a decrease of the in-stent restenosis hazard.

3.2.2
Effect of Nanostructures on Orientation and Physiology of Osteoblasts

Thomas K. Monsees and Richard H. W. Funk

On a micrometer-sized patterned implant, adherent growing cells often migrate and align in the direction of the surface structure (contact guidance) [22]. Depending on the surface topography, the cells can finally form a well-organized structure. In contrast, random cell orientations are often found on smooth surfaces. This may lead to increased scar tissue formation during wound healing [23]. Surface structures on a micrometer scale may also affect cell shape by elongating them along the grooves, which in turn can influence cell proliferation, differentiation and function [24–27]. Recently, the influence of nanostructures on cell behavior was also shown: Teixeira *et al.* [28] found that corneal epithelial cells elongate and align along silicone ridges of 70 nm width and 150 nm depth. Reduced fibroblast adhesion was observed using nanosized pits on polymethylmethacrylate (35 to 120 nm diameter) [29,30]. On 13 nm high polymer islands, fibroblasts showed increased cell spreading and up-regulation of gene expression [31].

To analyze the effect of nanostructures on cell orientation and function, we used patterns of parallel titanium oxide lines. These structures have different widths (0.2–10 µm) and distances (2–20 µm, 1000 µm), but a common height of only 12 nm (see Section 3.4 of Part I for details).

Materials and Methods
Primary osteoblastic cells were isolated from fetal rat calvariae as described previously [32] and cultured in Ham's F12 medium containing 12% fetal calf serum. Human osteogenic sarcoma cells (SaOS-2; ATCC HTB 85) were cultured in McCoy's 5A medium containing 15% fetal calf serum. Sterilized titanium discs were placed into culture wells and carefully covered with either calvarial or SaOS-2 cells at a density of 10 000 cells cm^{-2}. This concentration was very low to exclude possible influences on cellular orientation by contact with other cells. Cell orientation was analyzed 48 h after plating. Cells were fixed and fluorescent-stained for actin and vinculin as described in Section 1.2. After fluorescent imaging, the cells were additionally stained with hematoxylin and eosin and analyzed using phase contrast microscopy. The proportion of cells with an alignment of their longitudinal axes in

the direction of the titanium oxide lines (orientation angle 0–30°, parallel; 60–90°, vertical) was considered to show contact guidance. Randomly orientated cells (45° angle or round or triangular shape) were not considered. Cells were fixed with formalin, critical-point dried and sputtered with gold before SEM imaging (M. Gelinsky, Department of Material Sciences, TU Dresden). For determination of alkaline phosphatase (ALP) activity as differentiation marker, cells were stimulated with $50\,\mu g\,mL^{-1}$ ascorbic acid and 10 mM β-glycerophosphate starting at day 3 of culture. ALP activity was determined using either the fluorescence-based ELF-97 endogenous phosphatase detection kit (Molecular Probes) or the release of *p*-nitrophenolate from *p*-nitrophenylphosphate at pH = 10.0 [32].

Results

Table 3.1 shows the time course of ALP activity in calvaria osteoblasts grown on smooth or structured titanium (parallel titanium oxide lines, width 0.2 µm, distance 2 µm, height 12 nm) during the first 9 days post-plating. Specific ALP activity increased with time on both surfaces, but at the indicated time points no significant differences were observed. Also SaOS-2 cells grown for 8 d on the structured titanium (parallel titanium oxide lines with 2 µm distance) (Figure 3.19B) showed no major differences in ALP activity compared to smooth titanium (Figure 3.19A). The fluorescence staining intensity and the number of positive cells were similar in cells seeded on smooth or nanosize structured surfaces.

Regarding contact guiding, we found that SaOS-2 cells stretched their cytoskeleton to align along the parallel structures and formed small filopodia to make contact with the oxide lines (Figure 3.20A–C). In contrast, cells grown on smooth titanium did not show any preferences in cell alignment or form (Figure 3.20D). In the range of 2 to 20 µm, the proportion of aligned cells did not significantly depend on the distance of the titanium oxide lines (Table 3.2). Nevertheless, there was a trend to more orientated cells with the narrowest line distance. On average, 43% of the SaOS-2 cells orientated parallel to the oxide lines, whereas 34% orientated vertically. This difference was significant ($P = 0.03$) compared to the control (line distance 1000 µm). Here the cells were randomly distributed (Table 3.2). In conclusion, we found that a

Table 3.1 Effect of titanium surface structure on the time course of alkaline phosphatase (ALP) activity in calvarial osteoblasts.

	µM pNPP/mg protein		
	5 days	**7 days**	**9 days**
Smooth	88.43 ± 6.00	110.65 ± 16.18	108.29 ± 16.01
Structured	90.50 ± 6.03	108.56 ± 5.69	131.68 ± 6.08

Cells were seeded on smooth or structured (parallel TiO$_2$ lines, width 0.2 µm, distance 2 µm, height 12 nm) titanium. ALP activity was measured by determining *p*-nitrophenolate (pNPP) released from *p*-nitrophenylphosphate at 405 nm. Results are presented as mean ± standard deviation ($n = 3$).

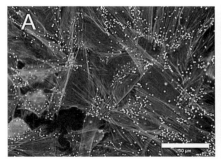

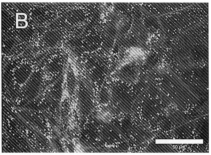

Figure 3.19 Influence of titanium surface structure on alkaline phosphatase (ALP) activity in SaOS-2 osteoblast-like cells. ALP activity was detected by immune histochemical staining and fluorescence microscopy. Cells were seeded at a density of 10 000 cells cm^{-2} onto either smooth (A) or structured (B, parallel TiO$_2$ lines, distance 2 μm, width 0.2 μm, height 12 nm) titanium discs. At day 8 of culture, cells were fixed and fluorescence-stained for actin (phalloidin-TRITC, fibres) and ALP activity (ELF-97 immunohistochemistry kit, molecular probes, dots). Scale bar = 50 μm.

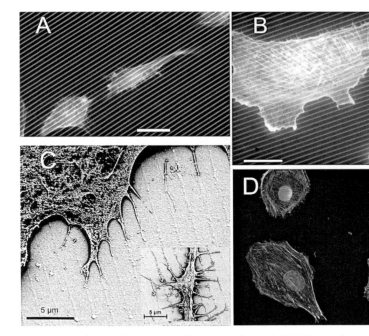

Figure 3.20 Influence of titanium surface structure on the alignment of osteoblast-like cells. SaOS-2 cells were analyzed two days after plating on (A–C) structured (parallel titanium oxide lines, height 12 nm, width 0.7 μm, distance 5 μm) or (D) smooth titanium sur- faces. (A, B) Cells were fluorescence-stained for actin, titanium oxide lines were visualized by brightfield microscopy and appeared in white. Scale bar = 20 μm. (C) Scanning electron microscopy image. (D) Cells were fluorescence-stained for actin and vinculin.

Table 3.2 Influence of titanium surface structure on orientation of SaOS-2 osteoblast-like cells.

Line distance (µm)	Line width (µm)	Vertical (%)	Parallel (%)	Random (%)	Microscopic fields
2	0.2	$34.2 \pm 8.6^{a,b)}$	$46.0 \pm 7.7^{c)}$	19.6 ± 3.2	19
5	0.7	$33.8 \pm 6.2^{a,b)}$	$43.5 \pm 10.7^{c)}$	22.5 ± 7.3	14
10	1.2	$35.9 \pm 10.1^{a,b)}$	$42.4 \pm 10.0^{c)}$	21.5 ± 7.3	8
20	1.5	$32.5 \pm 23.4^{a,b)}$	$41.9 \pm 14.2^{c)}$	25.6 ± 23.4	4
Average, 2–20		$34.1 \pm 12.0^{d)}$	$43.4 \pm 10.6^{d)*}$	22.3 ± 10.3	45
Control, 1000	10.0	34.2 ± 13.2	34.6 ± 11.5	32.5 ± 11.0	19

Values are mean ± standard deviation.
*$P < 0.05$.
a) Versus parallel for a specific line distance.
b) Versus vertical control.
c) Versus parallel control.
d) Versus control.

significantly increased proportion of 9.3% of the SaOS-2 cells prefer a more parallel rather than a vertical alignment on titanium oxide lines with a distance of 2–20 µm.

Discussion
We found no significant differences in specific ALP activities in calvarial osteoblasts or SaOS-2 cells grown on nanoscale-structured titanium implants compared to those grown on smooth titanium [32]. We also noticed no variations in corresponding ALP gene expression as determined by RT-PCR (not shown). A structured implant, as used in this study, has a rough surface. Our finding is in line with data of Rosa and Beloti [33], who found no significant effects of titanium or Ti6Al4V surface roughness ($R_a = 0.24–1.98$ µm) on ALP activity in rat bone marrow cells. In contrast, other studies showed increasing ALP activities of different cell types on rough titanium or Ti6Al4V surfaces [34,35]. Differences in cell origin, osteoblastic differentiation and especially design of the structured surfaces may explain these discrepancies.

However, we observed a significant amount of SaOS-2 cells orientated parallel with the titanium oxide structures. The cell shape aligned to the surface pattern with the edges lying on top of the lines. In the control pattern, however, with a line distance far beyond cell size or on the smooth titanium surface, cells distributed completely at random. The percentage of aligned cells did not significantly change with the distance of the titanium oxide lines within the range of 2 to 20 µm. Similar data were reported by Lehnert et al. [36]. Using extracellular matrix-coated dots, they showed that with increasing distance (5–20 µm) cells adapted their shape to the dot pattern and displayed straight edges from dot to dot. In contrast, a dot distance exceeding 30 µm limited cell spreading to triangular, ellipsoid or round forms. At distances less than 2 µm, cell morphology was equal to that on a homogeneous substrate. Also Teixeira et al. [28] found the percentage of aligned corneal epithelial cells to be constant on silicone patterns ranging from 0.4 to 2 µm.

The portion of cells aligning to a structured surface depends on the peak to valley height, with more orientated cells on the higher structures or the deeper grooves.

The oxide lines used in our experiments were only 12 nm in height, which is very low compared to the micrometer-sized groove and ridge pattern used in other studies [25–27]. Because the majority of the cells showed a parallel alignment, they may recognize the nanosize difference in height between titanium bulk surface and titanium oxide lines. Our results are supported by recent studies showing that mammalian cells indeed react to nanosize order and symmetry. Contact guidance on nanometer-sized depth grooves (150 nm and 100 nm) was observed using silicone and quartz as substrates, respectively [28,37]. Recently, Dalby et al. [38] demonstrated that fibroblasts can even "sense" structures as low as 10 nm in height.

Because of the different physiochemical properties at the interface of titanium bulk and laser-induced titanium oxide lines, cues other than contact guidance (e.g. differences in the electric potential or surfaces charge) may also affect orientation and migration of the osteoblasts. This topic is discussed in Section 2.1.1.

3.2.3
Cell Response to Ti6Al4V Modified with TiO$_2$ and RGDS Peptides on Different Surface Roughness

Bettina Hoffmann, Frank Heidenau, Rainer Detsch, and Günter Ziegler

Osteoblast-like MC3T3-E1 cells (DSMZ, Deutsche Sammlung von Mikroorganismen und Zellkulturen GmbH, Germany) were cultured for 3 h (serum free), 48 h and 28 d on samples with micro- and nanostructured surface modifications. In each cell test, samples were incubated with 100 000 cells mL^{-1} in a 24-well tissue culture plate (Greiner, Germany). For each experiment, four samples were used and always compared with polystyrene (PS) as a reference.

Cell number, cell viability and cell morphology were determined on Ti6Al4V discs with different roughness values (see Section 3.1 of Part I) after a defined incubation period (3 h, 48 h and 28 d). In addition to examinations of uncoated Ti6Al4V discs, the same investigations were carried out on TiO$_2$-coated (see Section 4.1.1 of Part I) and TiO$_2$-coated + RGDS-immobilized (see Section 4.3.1 of Part I) samples. Also, discs with different roughness values, which had been produced by mechanical treatment, were investigated by cell testing. Information about cell adhesion was obtained from cell tests with an incubation time of 3 h (Figure 3.21, left). To determine the direct influence of the interface on osteoblast-like cells, the incubation was carried out in culture medium without serum. Furthermore, cells were cultured over 48 h (Figure 3.21, middle) and 28 d (Figure 3.21, right) on all discs. At different incubation times, a similar trend for cell number and cell viability could be recognized (Figure 3.21). On uncoated Ti6Al4V and TiO$_2$-coated discs, a change of cell parameters with roughness variations is noticeable. With decreasing R_a values, cell number and cell viability increase. Additionally, an increase in cell number and viability on the TiO$_2$-coated discs can be seen in comparison to the uncoated ones. With an additional RGDS immobilization, cell number and cell viability become independent of the roughness value. The absolute values lie in the range of those of

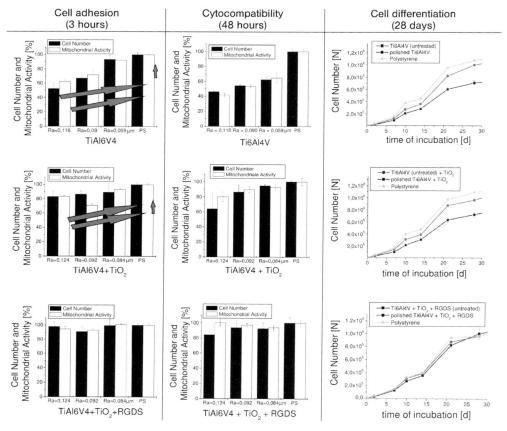

Figure 3.21 Cell tests on untreated Ti6Al4V (upper panel), TiO$_2$-coated Ti6Al4V (middle panel) and RGDS-functionalized TiO$_2$-coated Ti6Al4V (lower panel) with different incubation times.

the polished samples without RGDS, which show the highest cell number and viability of the TiO$_2$-coated discs.

Cell investigations on TiO$_2$-coated Ti6Al4V discs with additional TiO$_2$ powder show that a larger amount of TiO$_2$ powder results in a higher cell number and cell viability at an incubation time of 48 h. As with the mechanically treated discs, RGDS immobilization levels out the differences between TiO$_2$-filled coatings with different values of roughness. The changes in cell number and cell viability with roughness values, which can be seen both on mechanically treated and TiO$_2$-powder-filled coatings, are compensated by an additional RGDS immobilization.

Conclusion

The combined effect of topography (variations of roughness) and chemical surface treatments (TiO$_2$ coating and RGDS immobilization) was investigated. The cell number and mitochondrial activity show an increase with decreasing roughness on ground uncoated Ti6Al4V and TiO$_2$-coated Ti6Al4V discs at all determined incubation times (3 h up to 28 d). By additional immobilized RGDS surfaces the cell number and cell viability became independent of surface roughness. These results indicate that additional RGDS immobilization of these rough surfaces improves the cell attachment and compensates for the effect of surface roughness on cell proliferation and viability.

3.2.4
Endothelial Cell Morphology on Functionalized Microstructured Titanium

Tilo Pompe, Manuela Herklotz, and Carsten Werner

The strong correlation of substrate-dependent cell morphology and cell function is well known from several examples in the literature [39,40]. These show distinct effects on proliferation, apoptosis and differentiation by applying two-dimensional adhesive patterns for the growth of endothelial cells or mesenchymal stem cells. Furthermore, it is known that three-dimensional cell culture can strongly affect not only the phenotype of cellular adhesion sites, but also cell behavior [41,42]. From our own recent studies, we demonstrated that cellular adhesions and cell morphology can further be influenced by the physicochemistry of the substrate surface [43–45].

In this context we investigated endothelial cell behavior on three-dimensional microstructured titanium surfaces, which were coated with different maleic anhydride copolymers. By this approach we aimed to investigate the combined effect of three-dimensional micropattern and substrate physicochemistry on cell fate.

Materials and Methods
Micrometer-sized groves were prepared in silicon wafers by a lithographic process with sizes ranging from 5 to 80 μm (GeSiM, Großerkmannsdorf, Germany). They were 10 μm deep. A titanium surface was prepared by sputtering a 100 nm layer on top of these structures. By wet chemical surface preparation, thin films of different maleic anhydride copolymers (poly(octadecene-*alt*-maleic anhydride), poly(propene-*alt*-maleic anhydride), poly(ethylene-*alt*-maleic anhydride)) were covalently attached with aminosilanes to these surfaces (for details see [42,46]). By this procedure titanium surfaces were achieved with a variation of physicochemical interactions towards the bio-environment and a defined three-dimensional microstructure.

Prior to cell experiments the surfaces were coated with fibronectin (50 μg mL^{-1}). Then human umbilical vein endothelial cells were seeded for 5 days on top of these surfaces in endothelial cell growth medium (Promocell, Heidelberg, Germany) with medium change every second day.

For visualizing cell morphology by fluorescence microscopy (TCS SP1, Leica, Bensheim, Germany) cells were fixed and permeabilized and further stained with phalloidin-TRITC and anti-fibronectin antibodies (for details see [43]). Additional samples were fixed and prepared for scanning electron microscopy.

Results

The combinatorial variation of micrometer sized topography features of titanium substrates and surface physicochemistry can be used to strongly influence endothelial cell behavior. During the 5 days of cell culture the endothelial cells adapt their phenotype depending on the width of the microchannels and the interaction strength of extracellular matrix proteins, i.e. fibronectin, to the substrate surface. The matrix–substrate interaction was varied by different covalent surface coatings of thin films of maleic anhydride copolymers.

As shown in Figures 3.22 and 3.23, the cells grow in a flat and spread morphology on surfaces with high matrix–substrate interaction. For large channel width (>20 m) they can be found on the top of the microstructures as well as inside the channels. In contrast for narrow channels (<20 μm) the cells bridge the gaps and do not extend to the bottom.

For narrow channels a similar bridging is observed on surfaces with a weaker matrix–substrate interaction. However, for intermediate-sized channels (20–40 μm)

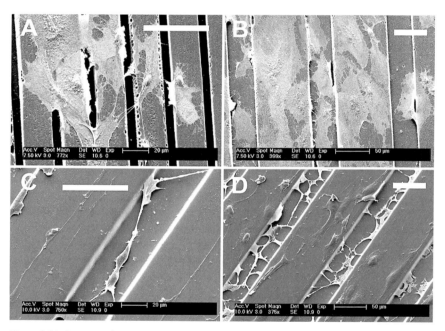

Figure 3.22 Scanning electron microscopy images of endothelial cells on functionalized titanium microstructures. (A) Small channel (<10 μm) shows, irrespective of their physicochemistry, an overgrowth with flat cell morphology. (B) On larger structures the cells still show a flat morphology on top and inside the channels, when the surface is functionalized with a polymer (poly(octadecene-*alt*-maleic anhydride)) showing strong interaction with fibronectin. (C, D) On intermediate-sized structures (20–40 μm) with a surface modification providing weak fibronectin–substrate interactions (poly(ethylene-*alt*-maleic anhydride)) cells are able to form multicellular assemblies with a connected extracellular matrix inside the channels. Scale bar: 50 μm.

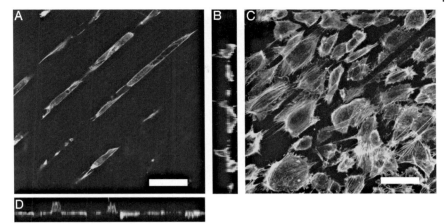

Figure 3.23 Laser scanning microscopy imaging visualizes the three-dimensional multicellular assemblies inside intermediate-sized channels with weak fibronectin–substrate interaction. The actin network was visualized by staining with phalloidin-TRITC. The main images (A, C) show horizontal section cuts through the image stack at half the height (A) and on top (C) rhead of a channel demonstrating in-growth of cells into the channels from the top cell layer. (B, D) Vertical sections through the image stack. Scale bar: 50 μm. Vertical height of section cuts (B, D): 16 μm.

the formation of extended elongated multicellular assemblies is observed pointing in the direction of vascular-like capillary formation. For larger microstructures the cells also start to exhibit a more elongated and thicker morphology on these surfaces; however, they do not establish many stable multicellular assemblies.

These results are summarized in Figure 3.24 demonstrating the combinatorial influence of microstructures and substrate–matrix interactions on the cellular morphology of endothelial cells. Starting of the formation of vascular-like structures

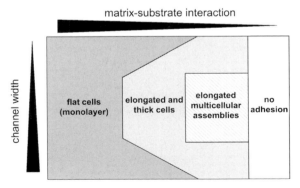

Figure 3.24 Schematic of the dependence of endothelial cell behavior on the size of titanium microstructures and matrix–substrate interaction, which is controlled by the physico-chemical surface properties of the polymer coatings.

is observed at an intermediate size of microchannels and intermediate interaction strength of extracellular matrix proteins to the substrate. Thus, this approach provides an idea as to how to influence endothelial cell phenotype in the context of vasculogenesis in tissue engineering technologies using artificial scaffolds. The range of useful microstructures and the application of a covalent surface modification by thin polymer films indicate further directions for scaffold design and functionalization.

3.3
Resistivity of Cells Under Shear Load

Andrei P. Sommer, Rosemarie Fuhrmann, and Ralf-Peter Franke

Evaluating the biocompatibility of a material involves studying its properties from various perspectives, including biomedical science, surface science, materials science and molecular biotechnology. Within a few milliseconds after an implant is inserted into the body, a biolayer consisting of water, proteins and other biomolecules from the physiological liquid is formed on the implant surfaces – this is known from the published literature. Subsequently, cells from the surrounding tissue migrate to the area around the implant due to stimulation by cytokines and growth factors in the biolayer. The interaction between an implant surface and cells is thus mediated trough the biolayer, in general. It becomes clear that the interface between biomaterial and cells is of central importance for defining and understanding biocompatibility.

For some years, biocompatibility has been defined as the ability of a material to perform with an appropriate host response in a specific application (Williams 1999). Under this definition, a material regarded as biocompatible in orthopedic surgery could be, for instance, inappropriate for cardiovascular applications. With the increasing number of artificial materials introduced into the field of biomedicine, there is an increasing demand for more discriminating tools to evaluate their biocompatibility. Any severe reactions by the host toward the biomaterial will probably result in failure. Consequently, the need for standardized methods and protocols for assessing the biological response of materials has never been greater. A list of tests being used to assess the biological response of materials can be found in ASTM.

Biocompatibility testing is sometimes described as a multitude of *in vitro* studies that are used in accordance with ISO 10993 to determine if a certain biomedical product is biocompatible. Clearly, *in vitro* tests cannot determine the biocompatibility of a material, at least for the moment. They provide vital information for extended decisions regarding animal testing, and finally clinical trials that will ultimately determine the degree of biocompatibility of the material in a specific application.

Present biocompatibility tests start with a variety of cell culture methods – studies that may assess the cytotoxicity, morphology or secretory functions of different cell types anticipated to contact the material that is targeted for implantation, as response to it. Other types of tests determine a catalogue of cell functions, for instance: adhesion, spreading, membrane integrity, replication, phagocytosis, the production of reactive oxygen species, secretion, activation, chemotaxis, chemokinesis and finally cell survival.

Blood contact assays have been developed and include tests investigating the adhesion or activation of blood cells, proteins and macromolecules such as those found in the complement or coagulation cascades. Other biocompatibility tests have been tentatively proposed and involve analytical testing or observations of physiological phenomena, reactions or surface properties assignable to a specific application such as antibacterial surface testing, protein adsorption characteristics, calcification or mineralization processes. The complexity in the classification of biocompatibility has been considerably simplified by discriminating between histocompatibility and hemocompatibility.

- Histocompatibility. Integration of body foreign materials with the proper short- and long-term function of the implant/tissue construct.
- Hemocompatibility. Biomaterials like cardiovascular implants, extracorporeal artificial organs and disposable medical devices often initiate coagulation processes, thrombocyte adhesion/activation and the immune response of the organism when they come into contact with blood. In this context, hemocompatibility describes the activation of platelets, the coagulation as the activation and polymerization of the fibrinogen–fibrin system, the complement activation, the activation of the bradykinin system, the activation of leukocytes with subsequent oxidative burst and proteolysis as well as the hemolysis describing the destruction of red blood cells.

Here we focus on physicochemically accessible determinants of histo- or hemo-compatibility, which comprise the microstructure, nanostructure, chemical composition, surface charge and water film structure [47–49]. These determinants can be assessed under conventional cell culture conditions or specifically under mechanical load, e.g., hydrodynamic shear stress. With the aforementioned five parameters, on the one hand, and the phenomenon that stem cells respond sensitively to these parameters, on the other hand, the possibility to examine stem cells under rheological stress [50] may offer us a realistic tool to separate stem cells from multiple differentiated cell populations (cell chromatography).

Methods

Based on our expertise regarding the rheological analysis of stem cells in general, and human bone marrow cells in particular, we investigated the possibility to specifically discriminate between stem cells and other cell populations (a multitude of more advanced bone marrow cells) via the rheometer established in our laboratory, which is shown in Figure 3.25.

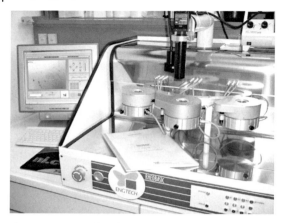

Figure 3.25 Cone-plate rheometer [47] to investigate cells suspended in culture medium under shear stress. Usual time of measurement: 3–5 h at low (0.5 dyn cm^{-2}) to medium (up to 3 dyn cm^{-2}) shear stress.

Results

Table 3.3 shows the results for six different substrate materials. Importantly, the nitrogen-doped titanium (TiN, N-rich) carried a significantly higher CD34 cell count than any other substrate material. This result cannot be understood without further examinations and hypotheses, in particular regarding specific (e.g. the "LIF-niche" as analogue for biological specificity to maintain mouse stem cell resting status) or nonspecific response of stem cells/progenitor cells.

Table 3.3 Substrate materials and test methods (dynamic, static) used to evaluate biological responses in cells. TiN stands for N-doped titanium, which can be stoichiometric (sto) or rich. Duration of shear load 3 h (at 0.5 dyn cm^{-2}).

Material	Culture condition	Total number (%)	CD34 positive (%)
Titanium	Static	100	20
TiN, N-sto	Static	17	29
TiN, N-rich	Static	32	50
Titanium	Dynamic	34	26
TiN, N-sto	Dynamic	17	29
TiN, N-rich	Dynamic	24	67

Discussion

It is not very possible that the TiN, N-rich result was due to a niche-specific stimulus. For this we feel justified in assuming that stem cells possess a capability to respond nonspecifically (to one or more of the five parameters mentioned above). In order to

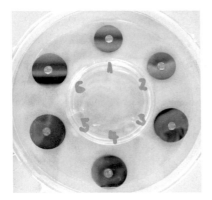

Figure 3.26 Drops of diluted aqueous nanosuspension of virtually equal volume on six different discs: titanium N-doped (stoichiometric) on glass (1), titanium N-rich on glass (2), titanium N-rich on Ti_6Al_4V (3), anatase on Ti_6Al_4V (4), rutile on Ti_6Al_4V (5) and titanium N-doped (stoichiometric) on Ti_6Al_4V (6).

identify the parameter responsible for the dominant effect, we applied drops of a diluted aqueous nanosuspension of virtually equal volume onto samples of the tested substrates and additionally on three related materials in order to establish a comparative scale. Figure 3.26 shows that despite the chemical contrast between the substrate materials there was no perceptible difference in the diameter (and because of virtually equal drop volumes in the contact angle) of the drops applied. The result is interesting, especially in view of the circumstance that there was no difference in the micro- and nanostructure between the samples numbered 3 and 6.

Conclusion

Interpreting the contact angle as a manifestation of the surface polarity, involving chemical and electrical properties, we are left with the explanation of the major differences in Table 3.3 with possible differences in the structure of the interfacial water layer at the substrates [47].

References

1 Nebe, B., Forster, C., Pommerenke, H., Behrend, D., Schmitz, K.P. and Rychly, J. (2001) Structural alterations of adhesion mediating components in cells cultured on poly-β-hydroxy butyric acid. *Biomaterials*, **22**, 2425–2434.

2 Anselme, K. (2000) Osteoblast adhesion on biomaterials. *Biomaterials*, **21**, 667–681.

3 Ezzell, R.M., Goldmann, W.H., Wang, N., Parasharama, N. and Ingber, D.E. (1997) Vinculin promotes cell spreading by mechanically coupling integrins to the cytoskeleton. *Exp. Cell Res.*, **231**, 14–26.

4 Wiesner, S., Legate, K.R. and Fässler, R. (2005) Integrin–actin interactions. *CMLS Cell Mol. Life Sci.*, **62**, 001–019.

5 Lange, R., Lüthen, F., Beck, U., Rychly, J., Baumann, A. and Nebe, B. (2002) Cell–extracellular matrix interaction and physico-chemical characteristics of titanium surfaces depend on the roughness of the material. *Biomol. Eng.*, **19**, 255–261.

6 Lüthen, F., Lange, R., Becker, P., Rychly, J., Beck, U. and Nebe, B. (2005) The influence of surface roughness of titanium on β1- and β3-integrin adhesion and the organization of fibronectin in human osteoblastic cells. *Biomaterials*, **26**, 2423–2440.

7 Diener, A., Nebe, B., Lüthen, F., Becker, P., Beck, U., Neumann, H.G. and Rychly, J. (2005) Control of focal adhesion dynamics by material surface characteristics. *Biomaterials*, **26**, 383–392.

8 Rychly, J., Pommerenke, H., Duerr, F., Schreiber, E. and Nebe, B. (1998) Analysis of spatial distributions of cellular molecules during mechanical stressing of cell surface receptors using confocal microscopy. *Cell Biol. Int.*, **22**, 7–12.

9 Nebe, B., Lüthen, F., Lange, R., Bulnheim, U., Müller, P., Neumann, H.G., Rychly, J. and Beck, U. (2005) Interface interaction of osteoblasts with structured titanium surfaces. *BIOmaterialien*, **6**, 35–41.

10 Nebe, B., Lüthen, F., Baumann, A., Diener, A., Beck, U., Neumann, H.G. and Rychly, J. (2003) Effects of titanium surface topography on the cell–extracellular matrix interaction in osteoblasts. *Mater. Sci. Forum*, **426–432**, 3023–3030.

11 Hale, L.V., Ma, Y.F. and Santerre, R.F. (2000) Semi-quantitative fluorescence analysis of calcein binding as a measurement of *in vitro* mineralization. *Calcif. Tissue Int.*, **67**, 80–84.

12 Nebe, B., Lüthen, F., Lange, R., Becker, P., Beck, U. and Rychly, J. (2004) Topography-induced alterations in adhesion structures affect mineralization in human osteoblasts on titanium. *Mater. Sci. Eng C*, **24**, 619–624.

13 Nebe, B., Bohn, W., Pommerenke, H. and Rychly, J. (1997) Flow cytometric detection of the association between cell surface receptors and the cytoskeleton. *Cytometry*, **28**, 66–73.

14 Schmidt, C., Pommerenke, H., Duerr, F., Nebe, B. and Rychly, J. (1998) Mechanical stressing of integrin receptors induces enhanced tyrosine phosphorylation of cytoskeletally anchored proteins. *J. Biol. Chem.*, **273**, 5081–5085.

15 Pankov, R., Cukierman, E., Katz, B.Z., Matsumoto, K., Lin, D.C., Lin, S., Hahn, C. and Yamada, K.M. (2000) Integrin dynamics and matrix assembly: tensin-dependent translocation of a5β1 integrins promotes early fibronectin fibrillogenesis. *J. Cell. Biol.*, **148**, 1075–1090.

16 Ingber, D.E. (1997) Tensegrity: the architectural basis of cellular mechanotransduction. *Annu. Rev. Physiol.*, **59**, 575–599.

17 Parker, K.K., Brock, A.L., Branwynne, C., Mannix, R.J., Wang, N., Ostuni, E., Geisse, N.A., Adams, J.C., Whitesides, G.M. and Ingber, D.E. (2002) Directional control of lamellipodia extension by constraining cell shape and orienting cell tractional forces. *FASEB J.*, **16**, 1195–1204.

18 Webster, T.J., Siegel, R.W. and Bizios, R. (1999) Osteoblast adhesion on nanophase ceramics. *Biomaterials*, **20**, 1231–1237.

19 Zhao, G., Zinger, O., Schwartz, Z., Wieland, M., Landolt, D. and Boyan, B.D. (2006) Osteoblast-like cells are sensitive to submicron-scale surface structure. *Clin. Oral Implants Res.*, **17**, 258–264.

20 Schwartz, Z., Nasazky, E. and Boyan, B.D. (2005) Surface microtopography regulates osteointegration: the role of implant surface microtopography in

osteointegration. *Alpha Omegan.*, **98**, 9–19.

21 Anselme, K., Bigerelle, M., Noel, B., Dufresne, E., Judas, D., Iost, A. and Hardouin, P. (2000) Qualitative and quantitative study of human osteoblast adhesion on materials with various surface roughnesses. *J. Biomed. Mater. Res.*, **49**, 155–166.

22 Brunette, D.M. (1986) Spreading and orientation of epithelial cells on grooved substrata. *Exp. Cell. Res.*, **167**, 203–217.

23 Wang, J.H.C., Grood, E.S., Florer, J. and Wenstrup, R. (2000) Alignment and proliferation of MC3T3-E1 osteoblasts in microgrooved silicone substrate subjected to cyclic stretching. *J. Biomech.*, **33**, 729–735.

24 Brunette, D.M. and Cheroudi, B. (1999) The effects of the surface topography of micromachined titanium substrata on cell behavior *in vitro* and *in vivo*. *J. Biomech. Eng.*, **121**, 49–57.

25 Eisenbarth, P., Linez, P., Biehl, V., Velten, D., Breme, J. and Hildebrand, H. (2002) Cell orientation and cytoskeleton organization on ground titanium surfaces. *Biomol. Eng.*, **19**, 233–237.

26 Soboyejo, W.O., Nemetski, B., Allameh, S., Marcantonio, N., Mercer, C. and Ricci, J. (2002) Interactions between MC3T3-E1 cells and textured Ti6Al4V surfaces. *J. Biomed. Mater. Res. A*, **62**, 56–72.

27 Lu, X. and Leng, Y. (2003) Quantitative analysis of osteoblast behavior on microgrooved hydroxyapatite and titanium substrata. *J. Biomed. Mater. Res. A*, **66**, 677–687.

28 Teixeira, I., Abrahams, G.A., Bertics, P.J., Murphy, C.J. and Nealey, P.F. (2003) Epithelial contact guidance on well-defined micro- and nanostructured substrates. *J. Cell Sci.*, **116**, 1881–1892.

29 Curtis, A.S., Gadegaard, N., Dalby, M.J., Riehle, M.O., Wilkinson, C.D. and Aitchison, G. (2004) Cells react to nanoscale order and symmetry in their surroundings. *IEEE Trans. Nanobiosci.*, **3**, 61–65.

30 Martines, E., McGhee, K., Wilkinson, C. and Curtis, A. (2004) A parallel-plate flow chamber to study initial cell adhesion on a nanofeatured surface. *IEEE Trans. Nanobiosci.*, **3**, 90–95.

31 Dalby, M.J., Yarwood, S.J., Riehle, M.O., Johnstone, H.J.H., Affrossman, S. and Curtis, A.S.G. (2002) Increasing fibroblast response to materials using nanotopography: morphological and genetic measurements of cell response to 13 nm high polymer demixed islands. *Exp. Cell Res.*, **276**, 1–9.

32 Monsees, T.K., Barth, K., Tippelt, S., Heidel, K., Gorbunov, A., Pompe, W. and Funk, R.H.W. (2005) Effects of different titanium alloys and nanosize surface patterning on adhesion, differentiation, and orientation of osteoblast-like cells. *Cells Tissues Organs*, **180**, 81–95.

33 Rosa, A.L. and Beloti, M.M. (2003) Rat bone marrow cell response to titanium and titanium alloy with different surface roughness. *Clin. Oral Impl. Res.*, **14**, 43–48.

34 Boyan, B.D., Batzer, R., Kiesewetter, K., Lie, Y., Cochran, D.L., Szmuckler-Moncler, S., Dean, D.D. and Schwartz, Z. (1998) Titanium surface roughness alters responsiveness of MG63 osteoblastic-like cells to $1\alpha,25\text{-}(OH)_2D_3$. *J. Biomed. Mater. Res. A*, **39**, 77–85.

35 De Santis, D., Guerriero, C., Nocini, P.F., Ungersbock, A., Richards, G., Gotte, P. and Armato, U. (1996) Adult human cells from jaw bones cultured on plasma-sprayed or polished surfaces of titanium or hydroxyapatite discs. *J. Mater. Sci. Mater. Med.*, **7**, 21–28.

36 Lehnert, D., Wehrle-Haller, B., David, C., Weiland, U., Ballestrem, C., Imhof, B.A. and Bastmeyer, M,. (2003) Cell behaviour on micropatterned substrata: limits of extracellular matrix geometry for spreading and adhesion. *J. Cell Sci.*, **117**, 41–52.

37 Clark, P., Connolly, P., Curtis, A.S., Dow, J.A. and Wilkinson, C.D. (1991)